Statistics For Dummies

Cheat Sheet

Common Statistics

The following are the most common statistics you're likely to come across, along with their formulas and a short description of what each one measures.

Statistic	Formula	Used For
Sample mean (average)	$\bar{x} = \dfrac{\sum x}{n}$	Weighted center of the data; affected by outliers
Median	Middle value in the ordered data set	Numerical center of the data; not affected by outliers
Sample standard deviation	$s = \sqrt{\dfrac{\sum (x - \bar{x})^2}{n-1}}$	Measure of variability; typical distance to the mean
Correlation coefficient	$r = \dfrac{1}{n-1} \sum \dfrac{(x - \bar{x})(y - \bar{y})}{s_x s_y}$	Strength and direction of linear relationship between x and y

Confidence Intervals

A confidence interval is an educated guess about some characteristic of the population (for example, what percentage of people in the United States own a cellphone?) A confidence interval contains an initial estimate (say, the average starting wage based on a sample of 1,000 recent graduates) plus or minus a margin of error (the amount by which you expect your results to vary, if a different sample were taken). The following are formulas for the most common confidence intervals. See Chapter 13 for details.

For	Statistic	Margin of Error	Use When
Population mean (μ)	$\bar{x}$	$\pm Z \times \dfrac{s}{\sqrt{n}}$	n is at least 30
Population mean (μ)	$\bar{x}$	$\pm t_{n-1} \times \dfrac{s}{\sqrt{n}}$	n is less than 30
Population proportion (p)	$\hat{p}$	$\pm Z \times \sqrt{\dfrac{\hat{p}(1 - \hat{p})}{n}}$	$n \times \hat{p}$ and $n(1 - \hat{p})$ are at least 5
Difference of two population means $(\mu_x - \mu_y)$	$(\bar{x} - \bar{y})$	$\pm Z \times \sqrt{\dfrac{s_1^2}{n_1} + \dfrac{s_2^2}{n_2}}$	n_1 and n_2 are both at least 30
Difference of two population proportions $(p_1 - p_2)$	$(\hat{p}_1 - \hat{p}_2)$	$\pm Z \times \sqrt{\dfrac{\hat{p}_1(1 - \hat{p}_1)}{n_1} + \dfrac{\hat{p}_2(1 - \hat{p}_2)}{n_2}}$	$1 \times \hat{p}$ and $n(1 - \hat{p})$ are at least 5 for each group

Confidence Coefficient (Z Values)

Confidence coefficients (Z-values) are an important component of confidence intervals. The Z-value is part of the margin of error — the amount you have to add or subtract in order to have a certain level of confidence in your results. To be more confident in your results, you need a larger Z-value. See Chapter 9 for details.

Confidence Level	Z-Value	Confidence Level	Z-Value
80%	1.28	95%	1.96
85%	1.44	98%	2.33
90%	1.64	99%	2.58

Statistics For Dummies®

Cheat Sheet

Hypothesis Tests

Hypothesis tests use data to test whether some claim about a population is true. (For example, someone may claim that 40% of Americans own a cellphone. Is that true?) To test a hypothesis, you take a sample, collect data, form a statistic, standardize it to form a *test statistic* (so it can be interpreted on a standard scale), and decide whether the test statistic supports the claim. (See Chapters 14 and 15 for more details.) The following are formulas for the most common hypothesis tests.

Test For	Null Hypothesis (Ho)	Test Statistic	Distribution	Use When
Population mean (μ)	$\mu = \mu_0$	$\dfrac{(\bar{x} - \mu_0)}{\frac{s}{\sqrt{n}}}$	Standard normal (Z)	n is at least 30
Population mean (μ)	$\mu = \mu_0$	$\dfrac{(\bar{x} - \mu_0)}{\frac{s}{\sqrt{n}}}$	t_{n-1}	n is less than 30
Population proportion (p)	$p = p_0$	$\dfrac{\hat{p} - p_0}{\sqrt{\frac{p_0(1 - p_0)}{n}}}$	Standard normal (Z)	$n \times p_0$ and $n(1 - p_0)$ are at least 5
Difference of two population means ($\mu_x - \mu_y$)	$\mu_x - \mu_y = 0$	$\dfrac{(\bar{x} - \bar{y}) - 0}{\sqrt{\frac{s_1^2}{n_1} + \frac{s_2^2}{n_2}}}$	Standard normal (Z)	n_1 and n_2 are both at least 30
Mean of difference (before − after)	$\mu_d = 0$	$\dfrac{\bar{d} - 0}{\frac{s}{\sqrt{n}}}$	Standard normal (Z)	30 or more pairs of data
Mean of difference (before − after)	$\mu_d = 0$	$\dfrac{\bar{d} - 0}{\frac{s}{\sqrt{n}}}$	t_{n-1}	Less than 30 pairs of data
Difference of two population proportions ($p_1 - p_2$)	$p_1 - p_2 = 0$	$\dfrac{(\hat{p}_1 - \hat{p}_2) - 0}{\sqrt{\hat{p}(1 - \hat{p})\left(\frac{1}{n_1} + \frac{1}{n_2}\right)}}$	Standard normal (Z)	$n \times \hat{p}$ and $n(1 - \hat{p})$ are at least 5 for each group

For Dummies: Bestselling Book Series for Beginners

Statistics

FOR

DUMMIES®

Statistics FOR DUMMIES®

by Deborah Rumsey, Ph.D.

WILEY

Wiley Publishing, Inc.

Statistics For Dummies®

Published by
Wiley Publishing, Inc.
111 River St.
Hoboken, NJ 07030
www.wiley.com

Copyright © 2003 by Wiley Publishing, Inc., Indianapolis, Indiana

Published by Wiley Publishing, Inc., Indianapolis, Indiana

Published simultaneously in Canada

For general information on our other products and services or to obtain technical support, please contact our Customer Care Department within the U.S. at 800-762-2974, outside the U.S. at 317-572-3993, or fax 317-572-4002.

Wiley also publishes its books in a variety of electronic formats. Some content that appears in print may not be available in electronic books.

Library of Congress Control Number: 2002102342

ISBN: 0-7645-5423-9

Manufactured in the United States of America

10 9 8 7

About the Author

Deborah Rumsey earned her Ph.D. in statistics from Ohio State University (OSU) in 1993. Upon graduating, she joined the faculty in the Department of Statistics at Kansas State University, winning the distinguished Presidential Teaching Award and earning tenure and promotion in 1998. In 2000, she returned to OSU as the Director of the Mathematics and Statistics Learning Center, where she is today. Deb is the Editor of the "Teaching Bits" of the *Journal of Statistics Education;* she has also published papers and given professional presentations on the subject of statistics education, with a particular emphasis on statistical literacy (skills for understanding statistics in everyday life and the workplace) and immersive learning environments (environments that promote students' discovery of ideas on their own). Her passions include fishing, bird watching, and Ohio State Buckeye football (not necessarily in that order).

Dedication

To my husband, Eric, and my son, Clint Eric: You are my greatest teachers.

Author's Acknowledgments

I'd like to thank the people that made this book possible: Kathy Cox, for asking me to write the book I'd always dreamed of writing; Tere Drenth, for handling my dream with just the right touch to make sure it was finished on time and in the right format; Janet Dunn, for her thorough copyedit; John Gabrosek, Grand Valley State University, for his thorough technical review. Also, many thanks for the Composition Services department at Wiley Publishing, Inc., who created all of the equations in this book and managed the difficult layout issues beautifully.

To Peg Steigerwald, Mike O'Leary, and my colleague Jim Higgins: The many scribbling-on-a-napkin conversations we've had about statistics helped shape the way I think and talk about it. Thank you for making it real and for cheering me on. Thanks to Tony Barkauskas, UW-LaCrosse, the first and best professor I ever had, for inspiring me. I am so grateful to my friends in the Department of Statistics and the Mathematics and Statistics Learning Center at Ohio State who gave me constant support and encouragement. And a big thank you to my family for their love and faith in me.

Publisher's Acknowledgments

We're proud of this book; please send us your comments through our Dummies online registration form located at www.dummies.com/register/.

Some of the people who helped bring this book to market include the following:

Acquisitions, Editorial, and Media Development

Project Editor: Tere Drenth

Acquisitions Editor: Kathy Cox

Copy Editor: Janet S. Dunn, Ph.D.

Technical Editor: John Gabrosek, Ph.D.

Editorial Manager: Michelle Hacker

Editorial Assistant: Elizabeth Rea

Cartoons: Rich Tennant, www.the5thwave.com

Production

Project Coordinator: Nancee Reeves

Layout and Graphics: Seth Conley, Carrie Foster, Kristin McMullan, Jacque Schneider, Erin Zeltner

Proofreaders: John Tyler Connoley, Andy Hollandbeck, Carl William Pierce, TECHBOOKS Production Services

Indexer: John Griffitts

Publishing and Editorial for Consumer Dummies

Diane Graves Steele, Vice President and Publisher, Consumer Dummies

Joyce Pepple, Acquisitions Director, Consumer Dummies

Kristin A. Cocks, Product Development Director, Consumer Dummies

Michael Spring, Vice President and Publisher, Travel

Brice Gosnell, Associate Publisher, Travel

Kelly Regan, Editorial Director, Travel

Publishing for Technology Dummies

Andy Cummings, Vice President and Publisher, Dummies Technology/General User

Composition Services

Gerry Fahey, Vice President of Production Services

Debbie Stailey, Director of Composition Services

Contents at a Glance

Table of Contents

Part V: Guesstimating with Confidence 189

Chapter 11: The Business of Estimation: Interpreting and Evaluating Confidence Intervals 191

Chapter 12: Calculating Accurate Confidence Intervals 197

Chapter 13: Commonly Used Confidence Intervals: Formulas and Examples 205

Part VI: Putting a Claim to the (Hypothesis) Test 213

Chapter 14: Claims, Tests, and Conclusions 215

Introduction

The purpose of this book is to get you ready to sort through and evaluate the incredible amount of statistical information that comes to you on a daily basis. (You know the stuff: charts, graphs, tables, as well as headlines that talk about the results of the latest poll, survey, experiment, or other scientific study.) This book arms you with the ability to decipher and make important decisions about statistical results (for example, the results of the latest medical studies), being ever aware of the ways in which people can mislead you with statistics and how to handle them.

This book is chock-full of real examples from real sources that are relevant to your everyday life: from the latest medical breakthroughs, crime studies, and population trends to surveys on Internet dating, cellphone use, and the worst cars of the millennium. By reading the chapters in this book, you begin to understand how to use charts, graphs, and tables, and you also know how to examine the results of the latest polls, surveys, experiments, or other studies. You even find out how to use crickets to gauge temperature and how to win a bigger jackpot in the lottery.

You also get to enjoy poking a little fun at statisticians (who take themselves too seriously at times). And that's because you don't have to be a statistician to understand statistics.

About This Book

This book departs from traditional statistics texts, references, supplement books, and study guides in the following ways:

- ✔ Practical and intuitive explanations of statistical concepts, ideas, techniques, formulas, and calculations.

- ✔ Clear and concise step-by-step procedures that intuitively explain how to work through statistics problems.

- ✔ Interesting real-world examples relating to your everyday life and workplace.

- ✔ Upfront and honest answers to your questions like, "What does this really mean?" and "When and how I will ever use this?"

Conventions Used in This Book

I have three conventions that you should be aware of as you make your way through this book.

- ✔ **Definition of sample size (*n*):** When I refer to the size of a sample, I usually mean the number of individuals selected to participate in a survey, study, or experiment. (The notation for sample size is *n*.) Suppose, however, that 100 people were *selected* to participate in a survey, and only 80 of them *responded:* Which of these two numbers is *n:* 100 or 80? The convention I use is 80, the number of people who actually responded, and this number may be fewer than the number asked to participate. So, any time you see the phrase "sample size," think of it as the final number of individuals who participated in and provided information for the study.

- ✔ **Dual-use of the word "statistics":** In some situations, I refer to statistics as a subject of study, or as a field of research, so the word is a singular noun. For example, "Statistics is really quite an interesting subject." In other situations, I refer to statistics as the plural of statistic, in a numerical sense. For example, "The most common statistics are the mean and the standard deviation."

- ✔ **Use of the word "data":** You're probably unaware of the debate raging amongst statisticians about whether the word "data" should be singular ("data is . . .") or plural ("data are . . ."). It got so bad that one group of statisticians had to develop two different versions of a statistics T-shirt recently: "Messy Data Happens" and "Messy Data Happen." At the risk of offending some of my colleagues, I'm going to go with the plural version of the word data in this book, because the word "data" is and will always be a plural noun (at least that's what my editor tells me).

Foolish Assumptions

I don't assume that you've had any previous experience with statistics, other than the fact that you're a member of the general public who gets bombarded every day with statistics in the form of numbers, percents, charts, graphs, "statistically significant" results, "scientific" studies, polls, surveys, experiments, and so on.

What I do assume is that you can do some of the basic mathematical operations and understand some of the basic notation used in algebra, such as the variables *x* and *y*, summation signs, taking the square root, squaring a number, and so on. If you need to brush up your algebra skills, check out *Algebra For Dummies* by Mary Jane Sterling (Wiley Publishing, Inc.).

Keep in mind, however, that statistics is really quite different from math. Statistics is much more about the scientific method than anything else, determining research questions; designing studies and experiments; collecting data; organizing, summarizing, and analyzing the data; interpreting results; and drawing conclusions. In a nutshell, you're using data as evidence to answer interesting questions about the world. Math comes into play only in terms of calculating summary statistics and performing some of the analyses, but that's just a tiny part of what statistics is really about.

I don't want to mislead you: You do encounter formulas in this book, because statistics does involve a bit of number crunching. But don't let that worry you. I take you slowly and carefully through each step of any calculations you need to do. I also provide examples for you to work along with this book, so that you can become familiar and comfortable with the calculations and make them your own.

How This Book Is Organized

This book is organized into seven major parts that explore the main objectives of this book, along with a final part that offers quick top-ten references for you to use. Each part contains chapters that break down each major objective into understandable pieces.

Part 1: Vital Statistics about Statistics

This part helps you become aware of the quantity and quality of statistics you encounter every day, in your workplace and in the rest of your life. You also find out that a great deal of that statistical information is incorrect, either by accident or by design. You also take a first step toward becoming statistically savvy by recognizing some of the tools of the trade, developing an overview of statistics as a process for getting and interpreting information, and getting up to speed on some statistical jargon.

Part II: Number-Crunching Basics

This part helps you become more familiar and comfortable with *data displays* (otherwise known as charts, graphs, and so on). It also gives you tips on interpreting these charts and graphs, as well as spotting a misleading graph right off the bat. You also find out how to summarize data by using some of the more common statistics.

Part III: Determining the Odds

This part uncovers the basics of probability: how you use it, what you need to know, and what you're up against when playing games of chance. The bottom line? Probability and intuition don't always mix!

You find out how probability factors into your daily life and get to know some basic rules of probability. You also get the lowdown on gambling: how casinos work and why the house always expects to win in the long run.

Part IV: Wading through the Results

In this part, you understand the underpinnings that make statistics work, including sampling distributions, accuracy, margin of error, percentiles, and standard scores. You understand how to calculate and interpret two measures of relative standing: standard scores and percentiles. You also get the lowdown on what statisticians describe as the "crown jewel of all statistics" (the central limit theorem) and how much more easily you can interpret statistics because of it. Finally, you begin to understand how statisticians measure variability from sample to sample, and why that's so important. In this part, you also find out exactly what that commonly used term — margin of error — means.

Part V: Guesstimating with Confidence

This part focuses on how you can make a good estimate for a population average or proportion when you don't know what it is. (For example, the average number of hours adults spend watching TV per week or the percentage of people in the United States who have at least one bumper sticker on their cars.) You also find out how you can make a pretty good estimate with a relatively small sample (compared to the population size). You get a general look at confidence intervals, find out what you use them for, understand how they're formed, and get the lowdown on the basic elements of a confidence interval (an estimate plus or minus a margin of error). You also explore the factors that influence the size of a confidence interval (such as sample size) and discover formulas, step-by-step calculations, and examples for the most commonly used confidence intervals.

Part VI: Putting a Claim to the (Hypothesis) Test

This part is about the decision-making process and the huge role that statistics plays in it. It shows you how researchers (should) go about forming and testing their claims, and how you can evaluate their results to be sure that they did the statistics right and have credible conclusions. You also review step-by-step directions for carrying out the calculations for commonly used hypothesis tests and for interpreting the results properly.

Part VII: Statistical Studies: The Inside Scoop

This part gives an overview of surveys, experiments, observational studies, and quality-control processes. You find out what these studies do, how they are conducted, what their limitations are, and how to evaluate them to determine whether you should believe the results.

Part VIII: The Part of Tens

This quick and easy part shares ten criteria for a good survey and ten common ways that statistics are misused and abused by researchers, the media, and the public.

Appendix

One of the main goals of this book is to motivate and empower you to be a statistical detective, digging deeper to find the real information you need to make informed decisions about statistics that you encounter. The appendix contains all of the sources that I use in my examples throughout this book, in case you want to follow up on any of them.

Icons Used in This Book

Icons are used in this book to draw your attention to certain features that occur on a regular basis. Here's what they mean:

This icon refers to helpful hints, ideas, or shortcuts that you can use to save time. It also highlights alternative ways to think about a particular concept.

 This icon is reserved for particular ideas that I hope you'll remember long after you read this book.

 This icon refers to specific ways that researchers or the media can mislead you with statistics and tells you what you can do about it.

 This icon is a sure bet if you have a special interest in understanding the more technical aspects of statistical issues. You can skip this icon if you don't want to get into the gory details.

Where to Go from Here

This book is written in such a way that you can start anywhere and still be able to understand what's going on. So take a peek at the Table of Contents or the Index, look up the information that interests you, and flip to the page listed.

Or, if you aren't sure where you want to start, consider starting with Chapter 1 and reading your way straight through the book.

Part I
Vital Statistics about Statistics

The 5th Wave   By Rich Tennant

Don't worry, this time I'm playing the odds. Statistically speaking, two good-luck charms should double my chances of winning.

In this part . . .

When you turn on the TV or open a newspaper, you're bombarded with numbers, charts, graphs, and statistical results. From today's poll to the latest major medical breakthroughs, the numbers just keep coming. Yet much of the statistical information you're asked to consume is actually wrong by accident — or even by design. How is a person to know what to believe? By doing a lot of good detective work.

This part helps awaken the statistical sleuth that lies within you by exploring how statistics affect your everyday life and your job, how bad much of the information out there really is, and what you can do about it. This part also helps you get up to speed with some useful statistical jargon.

Chapter 1

The Statistics of Everyday Life

In This Chapter

▶ Encountering statistics in everyday life: what you see and how often you see it

▶ Discovering how statistics are used in the workplace

Today's society is completely taken over by numbers. Numbers appear everywhere you look, from billboards telling of the latest abortion statistics, to sports shows discussing the Las Vegas odds for the upcoming football game to the evening news, with stories focusing on crime rates, the expected life span of someone who eats junk food, and the president's approval rating. On a normal day, you can run into five, ten, or even twenty different statistics (with even more on Election Night). Just by reading a Sunday newspaper all the way through, you come across literally hundreds of statistics in reports, advertisements, and articles covering everything from soup (how much does an average person consume per year?) to nuts (how many nuts do you have to eat to increase your IQ?).

The purpose of this chapter is to show you how often statistics appear in your life and work and how statistics are presented to the general public. After reading this chapter, you begin to see just how often the media hits you with numbers and how important it is to be able to unravel what all those numbers mean. Because, like it or not, statistics are a big part of your life. So, if you can't beat 'em, and you don't want to join 'em, you should at least try to understand 'em.

Statistics and the Media Blitz: More Questions than Answers?

Open a newspaper and start looking for examples of articles and stories involving numbers. It doesn't take long before numbers begin to pile up. Readers are inundated with results of studies, announcements of breakthroughs, statistical reports, forecasts, projections, charts, graphs, and summaries. The extent to which statistics occur in the media is mind-boggling. You may not even be aware of how many times you're hit with numbers in today's information age. Here are just a few examples from one Sunday paper's worth of news. While you're reading this, you may find yourself getting nervous, wondering what you can and can't believe anymore. Relax! That's what this book is for, helping you sort out the good from the bad information. (Chapters 2 through 5 give you a great start.)

Probing popcorn problems

The first article I come across that deals with numbers is entitled, "Popcorn plant faces health probe." The subheading reads "Sick workers say flavoring chemicals caused lung problems." The article describes how the Centers for Disease Control (CDC) is expressing concern about a possible link between exposure to chemicals in microwave popcorn flavorings and some cases of fixed obstructive lung disease. Eight people from one popcorn plant alone contracted this lung disease, and four of them were awaiting lung transplants. According to the article, similar cases were reported at other popcorn factories. Now, you may be asking, "What about the folks who eat microwave popcorn?" According to the article, the CDC finds "no reason to believe that people who eat microwave popcorn have anything to fear." (Stay tuned.) They say that their next step is to evaluate employees more in-depth, including surveys to determine health and possible exposures to the said chemicals, checks of lung capacity, and detailed air samples. The question here is: How many cases of this lung disease constitute a real pattern, compared to mere chance or a statistical anomaly? (More about this in Chapter 14.)

Venturing into viruses

The second article I find discusses the most recent cyber attack — a worm-like virus that has made its way through the Internet, slowing down Web browsing and e-mail delivery across the world. How many computers were affected? The experts quoted in the article say that 39,000 computers were infected, affecting hundreds of thousands of other systems. How did they get that number? Wouldn't that be a hard number to get hold of? Did they check each computer out there to see whether it was affected? The fact that this article was written less than 24 hours after the attack would suggest that this number is a guess. Then why say 39,000 and not 40,000? To find out more on how to guesstimate with confidence (and how to evaluate someone else's numbers) see Chapter 11.

Comprehending crashes

Next in the paper appears an alert about the soaring number of motorcycle fatalities. Experts say that these fatalities are up more than 50% since 1997, and no one can figure out why. The statistics tell an interesting story. In 1997,

2,116 motorcyclists were killed; in 2001, the number was 3,181, as reported by the National Highway Traffic Safety Administration (NHTSA). In the article, many possible causes for the increased motorcycle death rate are discussed, including the fact that riders today tend to be older (the average age of motorcyclists killed in crashes increased from 29.3 years in 1990 to 36.3 years in 2001).

Bigger bikes are listed as another possibility. The engine size of an average motorcycle has increased almost 25% — from 769 cubic centimeters in 1990 to 959 cubic centimeters in 2001. Another possibility may be that some states are weakening their helmet laws. The experts quoted in the article say that a more comprehensive causation study is needed, but such a study probably won't be done because it would cost between 2 and 3 million dollars. One issue that is not addressed in the article is the number of people riding motorcycles in 2001, compared to the number of riders in1997. More people on the roads generally means more fatalities, if all the other factors remain the same. However, along with the article is a graph showing motorcycle deaths per 100 million vehicle miles traveled in the United States from 1997 to 2001; does that address the issue of more people on the roads? A bar graph is also included, comparing motorcycle deaths to deaths that occurred in other types of vehicles. This bar graphs shows that motorcycle deaths occur at a rate of 34.4 deaths per 100 million vehicle miles traveled, compared to just 1.7 deaths for the same number of miles traveled in cars. This article has lots of numbers and statistics, but what does it all mean? The number and types of statistics can quickly get confusing. Chapter 4 helps you sort out graphs and charts and the statistics that go along with them.

Mulling malpractice

Further along in the newspaper is a report about a recent medical malpractice insurance study, which may affect you in terms of the fees your doctor charges and your ability to get the health care you need. So what's the extent of the problem? The article indicates that 1 in 5 Georgia doctors has stopped doing risky procedures (like delivering babies) because of the ever-increasing malpractice insurance rates in the state. This is described as a "national epidemic" and a "health crisis" around the country. Some brief details of the study are included, and the article states that of the 2,200 Georgia doctors surveyed, 2,800 of them — which they say represents about 18% of those sampled — were expected to stop providing high risk procedures. Wait a minute! Can that be right? Out of 2,200 doctors, 2,800 don't perform the procedures, and that is supposed to represent 18%? That's impossible! You can't have a bigger number on the top of a fraction, and still have the fraction be under 100%, right? This is one of many examples of errors in statistics that are reported in the media. So what's the real percentage? You can only guess. Chapter 5 nails down the particulars of calculating statistics, so that you can know what to look for and immediately tell when something's not right.

Belaboring the loss of land

In the same Sunday paper is an article about the extent of land development and speculation across the country. Given the number of homes likely being built in your neck of the woods, this is an important issue to get a handle on. Statistics are given regarding the number of acres of farmland that are being lost to development each year and also translates those acres to square miles. To further illustrate how much land is being lost, the area is also listed in terms of the number of football fields. In this particular example, experts say that the mid-Ohio area is losing 150,000 acres per year, which is 234 square miles, or 115,385 football fields (including end zones). How do people come up with these numbers, and how accurate are they? And does it help to visualize land loss in terms of the corresponding number of football fields?

Scrutinizing schools

The next topic in the paper is school proficiency, specifically whether extra school sessions are helping students perform better. The article states that 81.3% of students in this particular district who attended extra sessions passed the writing proficiency test, while only 71.7% of those who didn't participate in the extra school sessions passed the proficiency test. But is this enough of a difference to account for the $386,000 price tag per year? And what's happening in these sessions to account for an improvement? Are students in these sessions spending more time just preparing for those exams, rather than learning more about writing in general? And here's the big question: Were those who participated in these extra sessions student volunteers who may be more motivated than the average student to try to improve their test scores? No one knows. Studies like this are going on all the time, and the only way to know what to believe is to understand what questions to ask, and to be able to critique the quality of the study. That's all part of statistics! The good news is, with a few clarifying questions, you can quickly critique statistical studies and their results. Chapter 17 helps you to do just that.

Trying to win the big one

Do you ever imagine winning the Super Lotto, a 1 in 89 million chance, on average?

Don't hold your breath! To put 1 in 89 million into perspective, imagine 89 million lottery tickets in one giant pile, with yours among them somewhere. Suppose I said that you have one chance to reach into the pile and pull out your own ticket — do you think you could do it? That is the same as your chance of winning one of those big lotteries. But with a bit of insider information, you can increase your jackpot if you do win. (I'd like a cut of your winnings if this turns out to work for you.) For more information on this and other gambling tips, see Chapter 7.

REMEMBER

Studying surveys of all shapes and sizes

Surveys and polls are probably the biggest vehicle used by today's media to grab your attention. It seems that everyone wants to do a survey, including market managers, insurance companies, TV stations, community groups, and even students in high-school classes. Here are just a few examples of survey results that are part of today's news.

With the aging of the American work force, companies are planning for their future leadership. (How do they know that the American workforce is aging, and if it is, by how much is it aging?) A recent survey shows that nearly 67% of human resources managers polled said that planning for succession had become more important in the past five years than it had been in the past. Now if you're thinking you want to quit your day job and apply to be a CEO, hold on. The survey also says that 88% of the 210 respondents said they usually or often fill senior positions with internal candidates. (But how many

managers did *not* respond, and is 210 respondents really enough people to warrant a story on the front page of the business section?) Believe it or not, when you start looking for them, you find numerous examples in the news of surveys based on far fewer participants than this.

Some surveys are based on lighter fare. For example, which device do Americans find most crucial today, their toothbrushes, bread machines, computers, cars, or cellphones? In a survey of 1,042 adults and 400 teens (how did they decide on those numbers?) 42% of adults and 34% of teens ranked the toothbrush as more important to them than cars, computers, or cellphones. Is this really big news? Since when should something as critical to daily hygiene as a toothbrush be lumped in with cellphones and bread machines? (The car came in second. But did you really need a survey to tell you that?) For more information on surveys, see Chapter 16.

Studying sports

The sports section is probably the most numerically jam-packed section of the newspaper. Besides the scores of the last game, the win/lose percentages for each team in the league, and the relative standing for each team, the specialized statistics reported in the sports world are so thick that they require wading boots to get through. For example, the basketball statistics are broken down by team, by quarter, and even by player. And you need to be a basketball junkie to interpret all of this, because everything is abbreviated (with no legend provided if you're out of the loop):

- ✔ MIN: Minutes played
- ✔ FG: Field goals
- ✔ FT: Free throws
- ✔ RB: Rebounds
- ✔ A: Assists

> ✔ PF: Personal fouls
>
> ✔ TO: Turnovers
>
> ✔ B: Blocks
>
> ✔ S: Steals
>
> ✔ TP: Total points

Who needs to know this, besides the players' mothers? Statistics are something that sports fans can never get enough of and that players can't stand to hear about. Stats are the substance of water-cooler debates and the fuel for armchair quarterbacks around the world.

Banking on business news

In the business section of the newspaper, you find statistics about the stock market. It was a bad week last week, with the stock market going down 455 points; is that decrease a lot or a little? You need to calculate a percentage to really get a handle on that. In the same business section, you also find reports on the highest yields nationwide on every kind of CD imaginable. (By the way, how do they know they're the highest?) You also see reports about loan rates: rates on 30-year fixed loans, 15-year fixed loans, 1-year adjustable rate loans, new car loans, used car loans, home equity loans, and loans from your grandmother (well actually no, but if grandma knew how to read these statistics, she may consider increasing the cushy rates she lets you have on her money!). Finally, you see numerous ads for those beloved credit cards — ads listing the interest rates, the annual fees, and the number of days in the billing cycle for the credit cards. How do you compare all of the information about investments, loans, and credit cards in order to make a good decision? What statistics are most important? The real question is, are the numbers reported in the paper giving the whole story, or do you need to do more detective work to get at the truth? Chapter 3 helps you start tearing apart these numbers and making decisions about them.

Taking in the travel news

You can't even escape the barrage of numbers by escaping to the travel section. In that section, I find that the most frequently asked question coming in to the Transportation Security Administration's response center (which receives about 2,000 telephone calls, 2,500 e-mail messages, and 200 letters per week on average — would you want to be the one counting all of those?) is, "Can I carry this on a plane?" where "this" can refer to anything from an animal to a giant tin of popcorn. (I wouldn't recommend the tin of popcorn.

You have to put it in the overhead compartment horizontally, and because things shift during flight, the cover will likely open; and when you go to claim your tin at the end of the flight, you and your seatmates will be showered. Yes, I saw it happen once.)

This leads to an interesting statistical question: How many operators will you need at various times of the day to field those calls that will come in? Estimating the number of anticipated calls is your first step, and being wrong can cost you money (if you overestimated it) or a lot of bad PR (if you underestimated it).

Talking sex (and statistics) with Dr. Ruth

On the accent page of the Sunday paper, you can read about Dr. Ruth's latest research on people's sex lives. She reports that sex doesn't stop at age 60 or even age 70. That's nice to know, but how did she determine this, and to what extent are people having sex at these ages? She doesn't say (maybe some statistics are better left unsaid, huh?). However, Dr. Ruth does recommend that folks in this age group disregard the surveys that report how many times a week, month, or year a couple has sex. In her view, this is just people bragging. She may be right about this. Think about it, if someone conducted a survey by calling people on the phone asking for a few minutes of their time to discuss their sex lives, who is going to be the most likely to want to talk about it? And what are they going to say in response to the question, "How many times a week do you have sex?" Are they going to report something that is the honest truth, or are they going to exaggerate a little? Self-reported surveys can be a real source of bias, and can lead to misleading statistics. So, don't be too hard on Dr. Ruth (who, by the way, is the author of *Sex For Dummies,* 2nd Edition, published by Wiley Publishing, Inc.). How would you recommend she go about finding out more about this very personal subject? Sometimes, research is more difficult than it seems. Chapter 2 has more examples of how statistics can go wrong, and what to look for.

Whetting your appetite for weather

The weather report provides another mass of statistics, with its forecasts of the next day's high and low temperatures (how do they decide on 16 degrees and not 15 degrees?) and reports of the day's UV factor, pollen count, pollution standard index, and water quality and quantity. (How do they get these numbers, by taking samples? How many samples do they take, and where do they take them?) You can even get a forecast looking ahead 3 days, a week, or even a month or a year! How accurate are weather reports these days? Given the number of times you get rained on when they told you it was going to be sunny, you could say they still have work to do on those forecasts!

The Las Vegas odds

When looking at how numbers are used (and abused) in everyday life, you can't ignore the world of sports betting, a multi-billion-dollar-a-year business that includes the casual bettor as well as the professional gambler and the compulsive gambler. What kinds of topics can you bet on? Pretty much anything that's got two outcomes. The crazy wagers that a person can make in Las Vegas have no limit (no pun intended).

Here is a sampling of some of the burning issues regarding the Super Bowl that one can wager on at a sports book (a betting place) in Las Vegas:

✔ Which team will have the most penalty yards?

✔ Which team will score last in the first half?

✔ Will a 2-point conversion be attempted?

✔ What will happen first, a score or a punt?

✔ What will the total net yards by both teams be (over 675 or under 675)?

✔ Will both teams make a 33-yard or longer field goal?

Hmm. Why not throw in the number of pounds of guacamole consumed by the Super Bowl TV viewers versus the number of blades of grass (or turf) on the field? Gamblers, start counting.

Probability and computer modeling do play an important role in forecasting today's weather, though, and are especially helpful regarding major events such as hurricanes, earthquakes, and volcano eruptions. Of course computers are only as smart as the people who program them, so scientists still have much work to do before tornados can be predicted before they even begin (wouldn't that be great, though?). For more on modeling and statistics, see Chapter 6.

Musing about movies

Moving on to the arts section of the newspaper, you see several ads for current movies. Each movie ad contains quotes from certain movie critics, some of which read, "Two thumbs up!," "The supreme adventure of our time," "Absolutely hilarious," or "One of the top ten films of the year!" Do you pay attention to the critics? How do you determine which movies to go to? Experts say that while the popularity of a movie may be affected by the critics' comments (good or bad) in the beginning of a film's run, word of mouth is the most important determinant of how well a film is going to do in the long run.

Studies also show that the more dramatic a movie is, the more popcorn is sold. Yes, the entertainment business even keeps tabs on how much crunching you do at the movies. How do they collect all of this information, and how does it impact the types of movies that are made? This, too, is part of statistics:

designing and carrying out studies to help pinpoint an audience and find out what they like, and then using the information to help guide the making of the product. So the next time someone with a clipboard asks if you have a minute, you may want to stand up and be counted.

Highlighting horoscopes

Those horoscopes: You read them, but do you believe them? Should you? Can people predict what will happen more often than just by chance? Statisticians have a way of finding out, by using something they call a hypothesis test (see Chapter 14). So far they haven't found anyone that is able to read minds, but people still keep trying!

Using Statistics at Work

Take a break from the Sunday newspaper, which you read in the comfort of your home, and move on to the daily grind of the workplace. If you're working for an accounting firm, of course numbers are part of your daily life. But what about nurses, portrait studio photographers, store managers, newspaper reporters, office workers, or even construction workers? Do numbers play a role if that's your job? You bet. This section gives you a few examples of how statistics creep into every workplace.

You don't have to go very far to see the tracks of statistics and how it weaves its way in and out of your life and your work. The secret is being able to determine what it all means and what you can believe, and to be able to make sound decisions based on the real story behind those numbers so that you can handle and even become accustomed to the statistics of everyday life.

Delivering babies — and information

Sue works as a nurse during the night shift in the labor and delivery unit at a university hospital. She has several patients that she has to take care of in a given evening, and she does her best to accommodate everyone. Her nursing manager has told her that each time she comes on shift she should identify herself to the patient, write her name on the whiteboard in the patient's room, and ask the patient whether she has any questions. Why does she do this? Because after each mother comes home from the hospital, she receives a phone call a few days later asking about the quality of care, what was missed, what the hospital can do to improve its service and quality of care, and what the hospital staff can do to ensure that the hospital is chosen more often than the other hospitals in town. Quality service is important, and for new moms staying in the hospital, with nurses coming and going every eight hours, knowing the names of their nurses is important, because this helps them get their questions answered in a timely manner. Sue's raises depend on her ability to follow through with the needs of new mothers.

Posing for pictures

Carol recently started her job as a photographer for a department store portrait studio; one of her strengths is working with babies. Based on the number of photos purchased by customers over the years, this store has found that people will buy more of the posed pictures than the natural-looking ones. As a result, the store managers will encourage their photographers to take posed shots.

A mother comes in with her baby and has a special request, "Could you please not pose my baby too deliberately? I just like his pictures to look natural." What does Carol say? "Can't do that, sorry. My raises are based on my ability to pose a child well." Wow! You can bet that the mother making the request is going to fill out that survey on quality service after this session — and not just to get $2.00 off her next sitting (if she ever comes back to that studio).

Poking through pizza data

Terry is a store manager at a local pizzeria that sells pizza by the slice. He is in charge of determining how many workers to have on staff at a given time, how many pizzas to make ahead of time to accommodate the demand, and how much cheese to order and grate, all with minimal waste of wages and ingredients. It's Friday night at 12 midnight and the place is dead. Terry has five workers left and has 5 pans of pizza he could throw in the oven, making about 40 slices of pizza. Should he send two of his workers home? Should he put more pizza in the oven or hold off? Terry knows what is most likely to happen because the store owner has been tracking the demand for weeks now, and he knows that every Friday night things slow down between 10 and 12, but then the bar crowd starts pouring in around midnight, and the crowd doesn't let up until the doors close at 2:30 in the morning. So Terry keeps the workers on, puts in the pizzas starting at 30 minute intervals from 12:00 on, and is rewarded with a good night money-wise, with satisfied customers and with a happy boss. For more information on how to make good estimates using statistics, see Chapter 11.

Working in the office of statistics

Take DJ, the administrative assistant for a computer company. How can statistics creep into her office workplace? Easy. Every office is filled with people who want to know answers to questions, and they want someone to "Crunch the numbers" to "Tell me what this means" to "Find out if anyone has any hard data on this" or to simply say, "Does this number make any sense?" They need to know everything from customer satisfaction figures to changes in inventory during the year; from the percentage of time employees spend on e-mail to the cost of supplies for the last three years. Every workplace is filled with statistics, and DJ's marketability and value as an employee could go up if she's the one the head honchos turn to for help. Every office needs a resident statistician — why not let it be you?

Chapter 2

Statistics Gone Wrong

. . . .

In This Chapter

▶ Examining the extent of statistics abuse

▶ Breaking down common statistics no-no's

▶ Feeling the impact of statistics gone wrong

. . . .

The numbers explosion can leave you feeling overwhelmed and confused (but with the help of this book you'll be able to understand many of the statistics you encounter in everyday life)! The purpose of this chapter is to bring another emotion to the surface: skepticism! Not skepticism like, "I can't believe anything anymore," but skepticism like, "Hmm, I wonder where that number came from." "Is this really true?" "I need to find out more information about this study before I believe these results." The media present you with many examples of statistics gone wrong, and after you find out how to spot these problems, you'll be well on your way to becoming more confident about statistics, ready to tackle the numbers explosion!

Taking Control: So Many Numbers, So Little Time

Statistics end up on your TV and in your newspaper as a result of a process. First, the researchers who study an issue generate results; this group of people is composed of pollsters, doctors, marketing researchers, government researchers, and other scientists. They are considered the *original sources* of the statistical information. After they get their results, these researchers want to tell people about it, so they typically put out either a press release or a journal article. Enter the journalists, who are considered the *media source* of the information. Journalists hunt for interesting press releases, sort through the journals, and basically search for the next headline. When reporters complete their stories, statistics are sent out to the public. This can happen through a variety of media: TV, newspaper, magazines, Web sites, newsletters, and so on. Now the information is ready to be taken in by the third group, the *consumers* of the information (you!). You and other consumers of information are the ones faced with the task of listening to and reading the information, sorting through it, and making decisions about it. And as you may have guessed, at any stage in the process of doing research, communicating results, or consuming information, errors can take place, either unintentionally or by design.

Detecting Errors, Exaggerations, and Just Plain Lies

Statistics can go wrong for many different reasons. First, a simple, honest error could have taken place. This can happen to anyone, right?

Other times, the error is a little more than a simple, honest mistake. In the heat of the moment, because someone feels strongly about a cause and because the numbers don't quite bear out the point that the researcher wants to make, statistics get tweaked, or, more commonly, they get exaggerated, either in terms of their values or in the way they're represented and discussed.

Finally, you may encounter situations in which the numbers have been completely fabricated and could not be repeatable by anyone because the results never happened. This is the worst-case scenario, and it does happen in the real world.

This section gives you tips to help you spot errors, exaggerations, and lies, along with some examples of each type of error that you, as an information consumer, may encounter.

Checking the math

The first thing you want to do when you come upon a statistic or the result of a statistical study is to ask the question, "Is this number correct?" Don't assume that it is! You may be surprised at the number of simple arithmetic errors that occur when statistics are collected, summarized, reported, or interpreted. Keep in mind that another type of error is an *error of omission* — information that is missing that would have made a big difference in terms of getting a handle on the real story behind the numbers. That makes the issue of correctness difficult to address, because you're lacking information to go on.

To spot arithmetic errors or omissions in statistics:

✔ Check to be sure everything adds up. In other words, do the percents in the pie chart add to 100? Do the number of people in each category add up to the total number surveyed?

✔ Double check even the most basic of calculations.

✔ Always look for a total, so that you can put the results into proper perspective. Ignore results based on tiny sample sizes.

Microwaving just doesn't add up

Many statistics break results down into groups, showing the percentage of people in each group who responded in a certain way regarding a particular question or demographic factor (such as age, gender, and so on). That's an effective way to report statistics, as long as all the percentages add up to 100%.

For example, *USA Today* reported the results of an opinion research study done for Tupperware regarding microwaving leftovers. The story reported that 28% of the people surveyed said they microwave leftovers almost daily, 43% said they microwave leftovers two to four times a week, and 15% said they do it once a week. Assuming that everyone should fit into these results somewhere, the percentages should add up to 100, or close to it. Quickly checking, the total is 28% + 43% + 15% = 86%. What happened to the other 14%? Who was left out? Where do they fall in the mix? No one knows. The statistics just aren't adding up.

Four out of five — really?

Another item you can check quickly is whether the total number of respondents is given. As a quick example, you may remember the Trident gum commercials which said that "Four out of five dentists surveyed recommend Trident gum for their patients who chew gum." This commercial is quite a few years old, but recently it has been revived in a funny series of new commercials asking, "What happened to that fifth dentist?" and then showing some incidents that might have happened to the fifth dentist that stopped him or her from pushing the "yes" button. But here is the real question: How many dentists were really surveyed? You don't know, because the survey doesn't tell you. You can't even check the fine print, because in the case of this type of advertising, none is required.

Why would knowing the total number of respondents make a difference? Because the reliability of a statistic is, in part, due to the amount of information that went into the statistic (as long as it was good and correct information). When the advertisers say "four out of five dentists," there may have actually been only five dentists surveyed. Now, maybe 5,000 were surveyed, and in that case, 4,000 of them recommended the gum. The point is, you don't know how many dentists actually recommended the gum unless you do more detective work to find out. In most cases, the burden is on you, the consumer, to find that information. Unless you know the total number of people who took part in the study, you can't get any perspective on how reliable this information could be.

Uncovering misleading statistics

Even when you uncover an error in a statistic, you may not be able to determine whether the error was an honest mistake, or if someone with an agenda was conveniently stretching the truth. By far the most common abuse of statistics is a subtle, yet effective, exaggeration of the truth. Even when the math checks out, the underlying statistics themselves can be misleading; they could be unfair, stretch the truth, or exaggerate the facts. Misleading statistics are harder to pinpoint than simple math errors, but they can have a huge impact on society, and, unfortunately, they occur all the time.

Crime statistics that don't pay

When spotting misleading statistics, you want to question the type of statistic used. Is it fair? Is it appropriate? Does it even make practical sense? If you're worried only about whether the numbers add up or that the calculations were correct, you could be missing a bigger error in that the statistic itself is measuring the wrong characteristic.

Crime statistics are a great example of how statistics are used to show two sides of a story, only one of which is really correct. Crime is often discussed in political debates, with one candidate (usually the incumbent) arguing that crime has gone down during his or her tenure, and the challenger often arguing that crime has gone up (giving the challenger something to criticize the incumbent for). How can two political candidates talk about crime going in two different directions? Assuming that the math is correct, how can this happen? Well, depending on the way that you measure crime, it would be possible to get either result. Table 2-1 shows the number of crimes in the United States reported by the FBI from 1987 to 1997.

Table 2-1	Number of Crimes in the U.S. (1987–1997)
Year	*Number of Crimes*
1987	13,508,700
1988	13,923,100
1989	14,251,400

Year	Number of Crimes
1990	14,475,600
1991	14,872,900
1992	14,438,200
1993	14,144,800
1994	13,989,500
1995	13,862,700
1996	13,493,900
1997	13,175,100

Is crime going up or down? It appears to be moving down in general, but you could look at these data in different ways and present these numbers in ways that make the trend look different. The big question is, do these data tell the whole story?

For example, compare 1987 to 1993. In 1987 an estimated 13,508,700 crimes took place in the United States, and in 1993, the total number of crimes was 14,144,800. It looks like crime went up during those six years. Imagine if you were a candidate making a challenge for the presidency; you could build a platform around this apparent increase in crime. And if you fast-forward to 1996, the total number of crimes in that year was estimated to be 13,493,900, which is only slightly less than the total number of crimes in 1987. So, was very much done to help curb crime during the nine-year period from 1987 to 1993? In addition, these numbers don't tell the whole story. Is the total number of crimes for a given year the most appropriate statistic to measure the extent of crime in the United States?

Another piece of important information has been left out of the story (and believe me, this happens more often than you may think)! Something else besides the number of crimes went up between 1987 and 1993: the population of the United States. The total population of the country should also play a role in the crime statistics, because when the number of people living in the country increases, you'd also expect the number of potential criminals and potential crime victims to increase. So, to put crime into perspective, you must account for the total number of people as well as the number of crimes. How is this done? The FBI reports a crime index, which is simply a crime rate. A *rate* is a ratio; it's the number of people or events that you're interested in, divided by the total number in the entire group.

The lowdown on ratios, rates, and percents

Statistics have a variety of different units in which they are expressed, and this variety can be confusing.

- A *ratio* is a fraction that divides two quantities. For example, "The ratio of girls to boys is 3 to 2" means that for every 3 girls, you find 2 boys. It doesn't mean that only 3 girls and 2 guys are in the group; ratios are expressed in lowest terms (simplified as small as possible). So you could have 300 girls and 200 guys; the ratio would still be 3 to 2.

- A *rate* is a ratio that reflects some quantity per a certain unit. For example, your car goes 60 miles per hour, or a neighborhood burglary rate is 3 burglaries per 1,000 homes.

- A *percentage* is a number between 0 and 100 that reflects a proportion of the whole. For example, a shirt is 10% off, or 35% of the population is in favor of a four-day work week. To convert from a percent to a decimal, divide by 100 or move the decimal over two places to the left. To remember this more easily, just remember that 100% is equal to 1, or 1.00, and to get from 100 to 1 you divide by 100 or move the decimal over 2 places to the left. (And just do the opposite to change from a decimal to a percent.)

Percentages can be used to determine how much a value increases or decreases, relatively speaking. Suppose the crimes in one city went up from 50 to 60, while the number of crimes in another city went up from 500 to 510. Both cities had an increase of 10 crimes, but for the first city, this difference is much larger, as a percentage of the total number of crimes. To find the percentage increase, take the "after" amount, minus the "before" amount and divide that result by the "before" amount. For the first city, this means crime went up by $(60 - 50) \div 50 = 10 \div 50 = 0.20$ or 20%. For the second city, this change reflects only a 2% increase, because $(510 - 500) \div 500 = 10 \div 500 = 0.02$ or 2%. To find percentage decrease, do the same steps. You'll just get a negative number, indicating a decrease.

Table 2-2 shows the estimated population of the U.S. for 1987–1997, along with the estimated number of crimes and the estimated *crime rates* (crimes per 100,000 people).

Table 2-2	Number of Crimes, Estimated Population Size, and Crime Rates in the U.S. (1987–1997)		
Year	*Number of Crimes*	*Estimated Population Size*	*Crime Rate (Per 100,000 People)*
1987	13,508,700	243,400,000	5,550.0
1988	13,923,100	245,807,000	5,664.2

Year	Number of Crimes	Estimated Population Size	Crime Rate (Per 100,000 People)
1989	14,251,400	248,239,000	5,741.0
1990	14,475,600	248,710,000	5,820.3
1991	14,872,900	252,177,000	5,897.8
1992	14,438,200	255,082,000	5,660.2
1993	14,144,800	257,908,000	5,484.4
1994	13,989,500	260,341,000	5,373.5
1995	13,862,700	262,755,000	5,275.9
1996	13,493,900	265,284,000	5,086.6
1997	13,175,100	267,637,000	4,922.7

Looking again at 1987 compared to 1993, you can see that the number of crimes increased from 13,508,700 in 1987 to 14,144,800 in 1993. (Note that this represents a 4.7% increase, because 14,144,800 – 13,508,700 equals 636,100, and if you divide this number by the original value, 13,508,700, you get 0.047, which is 4.7%.) So, looking at it this way, someone may report that crime went up 4.7% from 1987 to 1993. But this 4.7% represents an increase in the *total number* of crimes, not the number of crimes *per person,* or the number of crimes per 100,000 people. To find out how the number of crimes per 100,000 people changed over time, you need to calculate and compare the crime rates for 1987 and 1993. Here's how: $(5,484.4 – 5,550.0) \div 5,550.0 = -65.6 \div 5,550.0 = -0.012 = -1.2\%$. The crimes per 100,000 people (crime rate) actually *decreased* by 1.2%

Depending on how you spin the numbers, the results could be made to show opposite trends: that crime went up or down between 1987 and 1993. But now that you know the difference between the number of crimes and the crime rate, you know that some statistics should not simply be reported as the total number of events, but instead should be reported as rates (that is, the number of events divided by the number in the entire group).

Question the type of statistic that was used before you try to make sense of the results. Is it a fair and appropriate measurement? Is it an accurate way to portray the real story behind the data, or is there a better way?

The scale tells you a lotto!

Charts and graphs are good ways of showing clearly and quickly the point that you want to make, as long as the drawings are done correctly and fairly. And just to be clear, what's the difference between a chart and a graph? Not much: Statisticians use these terms quite interchangeably when talking about visual displays of statistical information. But a good rule is that if the picture shows bars or pies or other figures or shapes, you're looking at a chart. Otherwise, you are looking at a graph, which plots numbers as they change over time, or as they appear as pairs on the *(x,y)* plane. (More on that stuff in Chapter 18.)

Unfortunately, many times, the charts and graphs accompanying everyday statistics aren't done correctly and/or fairly, and you need to be on the look-out for problems. One of the most important elements to watch for is the way that the chart or graph is scaled. The *scale* of a graph is the quantity used to represent each tick mark on the axis of the graph. Do the tick marks increase by 10s, 20s, 100s, 1,000s, or what? The scale can make a big difference in terms of the way the graph or chart looks.

For example, the Kansas Lottery routinely shows its recent results from the Pick 3 Lottery. One of the statistics reported is the number of times each number (0 through 9) is drawn among the three winning numbers. Table 2-3 shows a chart of the number of times each number was drawn through March 15, 1997 (during 1,613 total Pick 3 games, for a total of 4,839 numbers drawn). Depending on how you choose to look at these results, you can again make the statistics appear to tell very different stories.

Table 2-3	Number of Times Each Number Was Drawn (Kansas Pick 3 Lottery, through 3/15/97)
Number Drawn	*Number of Times Drawn*
0	485
1	468
2	513
3	491
4	484
5	480
6	487
7	482
8	475
9	474

The way lotteries typically display results like those in Table 2-3 is shown in Figure 2-1. Notice that in this chart, it seems that the number 1 doesn't get drawn nearly as often (only 468 times) as number 2 does (513 times). The difference in the height of these two bars appears to be very large, exaggerating the difference in the number of times these two numbers were drawn. However, to put this in perspective, the actual difference here is 513 – 468 = 45, out of a total of 4,839 numbers drawn. In terms of the total number of individual numbers drawn, the difference between the number of times the number 1 and the number 2 are drawn is 45 ÷ 4,839 = 0.009, or only nine-tenths of one percent.

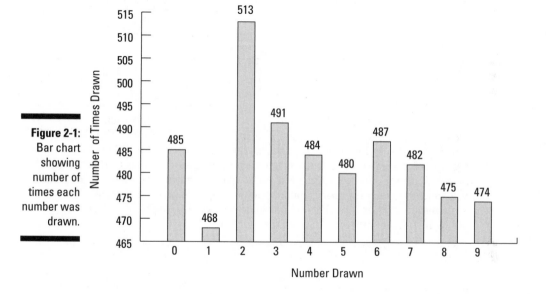

Figure 2-1: Bar chart showing number of times each number was drawn.

What makes this chart exaggerate the differences? Two issues come to the surface here, both affecting the appearance of the chart. First, notice that the vertical axis shows the number of times (or frequency) that each number is drawn, and it goes up by 5s. So a difference of 5 out of a total of 4,839 numbers drawn appears as if it actually means something. This is a common trick used to exaggerate results — stretching the scale so that differences appear larger than they really are. Second, the chart starts counting not at zero, but at 465, so it really is only showing the top part of each bar, where the differences are. This also exaggerates the results.

Table 2-4 shows a more realistic summary for each of the numbers drawn in the Pick 3 Lottery, by showing the percentage of times each number was drawn.

Table 2-4	Percentage of Times Each Number Was Drawn	
Number Drawn	*Number of Times Drawn*	*Percentage of Times Drawn*
0	485	10.0% = 485 ÷ 4,839
1	468	9.7% = 468 ÷ 4,839
2	513	10.6% = 513 ÷ 4,839
3	491	10.1% = 491 ÷ 4,839
4	484	10.0% = 484 ÷ 4,839
5	480	9.9% = 480 ÷ 4,839
6	487	10.0% = 487 ÷ 4,839
7	482	10.0% = 482 ÷ 4,839
8	475	9.8% = 475 ÷ 4,839
9	474	9.8% = 474 ÷ 4,839

Figure 2-2 is a bar chart showing the percentage of times each number was drawn, rather than the number of times each number was drawn. Note that this chart also uses a more realistic scale than the one in Figure 2-1, and that it also starts at zero, both of which make the differences appear as they really are — not much different at all. Boring, huh?

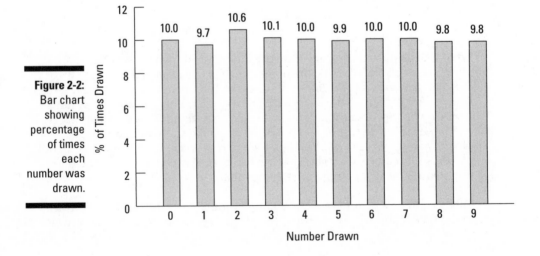

Figure 2-2:
Bar chart showing percentage of times each number was drawn.

Now why would a lottery do this? Maybe it wants you to believe you're getting some inside information, and thinking that the number 1 doesn't get drawn very much will make you want to buy a lottery ticket and choose 1, because it's "due" to happen (which is *not* true, by the way; see Chapter 7 for more on this). Or, you may want to choose the number 2, because it has been drawn a lot, and it's "on a roll" (again, no dice). However you look at it, the lottery folks want you to think that some "magic" is involved in the numbers, and you can't blame them; that's their business.

Misleading graphs occur all the time in the media! Reporters and others can stretch the scale out (make the tick marks represent increments of small amounts) and/or start at a number other than zero, to make differences appear larger than they really are. Or the scale can also be squeezed down (make the tick marks represent increments of large amounts) to give the appearance of "no change." These are examples of misleading representations of the truth. (See Chapter 4 for more information on spotting misleading graphs.)

Looking at the scale of a graph or chart can really help you keep the results in perspective.

Checking your sources

Check the source of the information; the best results are often published by a journal that's known by the experts in the field. For example, in the world of medical science, the *Journal of the American Medical Association* (JAMA), the *New England Journal of Medicine, The Lancet,* and the *British Medical Journal* are all among the reputable journals in which doctors publish results and read about new findings.

When examining the results of any study, consider the source and question all of the studies that were conducted, not just those whose results were published in journals or appeared in advertisements. A conflict of interest on the part of researchers can lead to incorrect information.

Counting on the sample size

Sample size isn't everything, but it does count for a great deal in terms of surveys and studies. If the study is designed and conducted correctly, and if the participants are selected randomly (that is, with no bias; see Chapter 3 for more on random samples), sample size is an important factor in determining the accuracy and repeatability of the results. (See Chapters 16 and 17 for more information on designing and carrying out studies.)

You may think that all studies are based on large numbers of participants. This is true for most surveys, but it isn't always true for other types of research, such as studies involving carefully controlled experiments. Experiments can be very time consuming; sometimes they take months or years to conduct in a variety of situations. Experimental studies can also be costly. Some experiments involve examining not people but products, such as computer chips or military equipment costing thousands or even millions of dollars. If the experiment involves destroying the product in the process of testing it, the cost of each experiment can be quite high. Because of the high cost of some types of research, some studies are based on a small number of participants or products. But fewer participants in a study (or fewer products tested) means less information overall, so studies with small numbers of participants (or products) in general are less accurate than similar studies with larger sample sizes.

Most researchers try to include the largest sample size they can afford, and they balance the cost of the sample size with the need for accuracy. Sometimes, though, people are just lazy and don't want to bother with a large enough sample. Sometimes, those researchers don't really understand the ramifications of having a small sample. And some folks hope you won't understand the importance of sample size, but now you do.

The worst examples of woefully inadequate sample size I've seen are TV ads where the sample size is only one. Usually, these commercials present what look like experiments to try to persuade the viewers that one product is superior to another. You've probably seen the TV commercial pitting one paper towel brand against another, where one piece of each type of paper towel is used to try to absorb the same amount of red juice. These examples may sound silly, but anyone can easily fall into the trap of drawing conclusions based on a sample size of one. (Have you ever told someone not to buy a product because you had a bad experience with it?) Remember that an anecdote (or story) is really a study with a sample size of only one.

Check the sample size to be sure the researchers have enough information on which to base their results. The margin of error (see Chapter 10) also gives you an idea of the sample size, because a small margin of error most likely means that a large sample was used.

Your doctor's time: Quantity or quality?

Headlines are the media's bread and butter, but headlines can also be misleading. Oftentimes, the headlines are more grandiose than the "real" information, especially when the stories involve statistics and the studies that generated the statistics. In fact, you'll often see a real gap between a headline and the "fine print" in such media stories.

A study conducted a few years back evaluated videotaped sessions of 1,265 patient appointments with 59 primary-care physicians and 6 surgeons in Colorado and Oregon. This study found that physicians who had *not* been sued for malpractice spent an average of 18 minutes with each patient, compared to 16 minutes for physicians who *had* been sued for malpractice.

Wow, is two minutes really that important? When the study was reported by the media with the headline, "Bedside manner fends off malpractice suits," this study seemed to say that if you are a doctor who gets sued, all you have to do is spend more time with your patients, and you're off the hook.

What's really going on? Am I supposed to believe that a doctor who has been sued needs only add a couple more minutes of time with each patient, and he can avoid being sued? Think about some of the other possibilities that may be involved here. It could be the case that doctors who don't get sued are just better doctors, ask more questions, listen more, and tell their patients more of what to expect, thereby taking more time; if so, what a doctor does during that time counts much more than how much time the doctor actually spends with each patient. But what about this possibility: Maybe the doctors who get sued are doing more difficult types of operations, or maybe they're specialists of some kind. Unfortunately, the article doesn't give you this information. Another possibility is that the doctors who don't get sued have fewer patients and, therefore, are able to spend more time with each patient and keep better track of them. At any rate, the fine print here doesn't quite match the hype, and when you read or hear about stories like these, watch out for similar gaps between what the headline claims and what the study actually found.

Reporting beyond the scope

You may wonder how political candidates find out about how their constituents feel. They conduct polls and surveys. Many surveys are done by an independent group, such as The Gallup Organization; others are done by representatives of the candidates themselves, and their methods can differ greatly from candidate to candidate, or from survey to survey.

In the 1992 presidential election, Ross Perot made quite a splash on the political scene. His group, United We Stand America, gained momentum and, ultimately, Ross Perot and his supporters made an impact on the election results. Often during debates and campaign speeches, Perot would give statistics and make conclusions about how Americans felt about certain issues. But was Mr. Perot always clear on how "Americans" felt, or was he simply clear about how his supporters felt? One of the vehicles Ross Perot used to get a handle on the opinions of Americans was to publish a questionnaire in the March 21, 1992, *TV Guide,* asking people to fill it out and send it to the address provided. Then he compiled the results of the survey and made them part of his campaign platform. From these results, he concluded that over 80% of the American people agreed with him on these issues. (Note, however, that he received only 18.91% of the vote in 1992.)

Part of the trouble with Mr. Perot's claims is the way the survey was conducted. In order to respond, you had to purchase the *TV Guide*, you had to have strong enough feelings about the survey to fill it out, and you had to mail it in yourself with your own stamp. Who is most likely to do this? Those who have strong opinions. In addition, the wording of the questions in this survey probably encouraged the people who agreed with Ross Perot to fill out the survey and send it in; those who didn't agree with him were more likely to ignore the survey.

If you can tell, based on the wording of the question, how the researcher wants you to respond to it, you know you're looking at a *leading question.* (See Chapter 16 for more information on how to spot this and other problems with surveys.)

Here is a sampling of some questions Mr. Perot used in his questionnaire. I paraphrased them, but the original intent is intact. (And this is not to pick on Mr. Perot; many political candidates and their supporters do the same type of thing.)

- ✔ Should the line-item veto be able to be used by the president to eliminate waste?

- ✔ Should Congress exclude itself from legislation it passes for us?

- ✔ Should major new programs be first presented to the American people in detail?

The opinions of the people who knew about the survey and chose to participate in it were more likely to be those who agreed with Mr. Perot. This is one example where the conclusions of a study went beyond the scope of the study, because the results didn't represent the opinions of "all Americans" as some voters were led to believe. How can you get the opinions of all Americans? You need to conduct a well-designed and well-implemented survey based on a random sample of individuals. (See Chapter 16 for more information about conducting a survey.)

When examining the conclusions of any study, look closely at both the group that was actually studied (or the group that actually participated) and the larger group of people (or lab mice, or fleas, depending on the study) that the studied group is supposed to represent. Then look at the conclusions that are made. See whether they match. If not, make sure you understand what the real conclusions are, and be realistic about the claims being made before you make any decisions for yourself.

Looking for lies in all the right places

You've seen examples of honest errors that lead to problems and how stretching, inflating, or exaggerating the truth can lead to trouble. Occasionally, you may also encounter situations in which statistics are simply made up, fabricated, or faked. This doesn't happen very often, thanks to peer-reviewed journals, oversight committees, and government rules and regulations.

But every once in a while, you hear about someone who faked his or her data, or "fudged the numbers." Probably the most commonly committed lie involving statistics and data is when people throw out data that don't fit their hypothesis, don't fit the pattern, or appear to be "outliers." In cases when someone has clearly made an error (for example, someone's age is recorded as being 200 years) it makes sense to try to clean up the data by either removing that erroneous data point or by trying to correct the error. But just because the data don't go your way, you can't just throw out some portion of it. Eliminating data (except in the case of a documented error) is ethically wrong; yet, it happens.

Regarding missing data from experiments, a commonly used phrase is "Among those who completed the study . . . ." What about those who didn't complete the study, especially a medical one? Did they die? Did they get tired of the side effects of the experimental drug and quit? Did they feel pressure to give certain answers or to conform to the researcher's hypothesis? Did they experience too much frustration with the length of the study and didn't feel they were getting any better, so they gave up?

Not everyone responds to surveys, and even people who generally try to take part in surveys sometimes find that they don't have the time or interest to respond to every single survey that they're bombarded with. American society today is survey-crazy, and hardly a month goes by when you aren't asked to do a phone survey, an Internet survey, or a mail survey on topics ranging from your product preferences to your opinion on the new dog-barking ordinance for the neighborhood. Survey results are only reported for the people who actually responded, and the opinions of those who chose to respond may be very different from the opinions of those who chose not to respond. Whether the researchers make a point of telling you this, though, is another matter.

For example, someone can say that he or she sent out 5,000 surveys, received 1,000 surveys back, and based the results on those responses. You may then think, "Wow, 1,000 responses. That's a lot of data; that must be a pretty accurate survey." Wrong. The problem is, 4,000 people who were selected to participate in the survey chose not to, and you have no idea what they would have said if they had responded. You have no guarantee that the opinions of these 4,000 people are represented by the folks who responded. In fact, the opposite could be true.

What constitutes a high *response rate* (that is, the number of respondents divided by the number of surveys sent out)? Some statisticians would settle for nothing less than 70%, but as TV's Dr. Phil would say, statisticians need to "get real." Rarely does a survey achieve that high of a response rate. But in general, the lower the response rate, the less credible the results and the more the results will favor the opinions of those who responded. (And keep in mind that respondents tend to have stronger opinions about the issues they choose to respond to.)

To watch for fake or missing data, look for information about the study including how many people were chosen to participate, how many finished the study, and what happened to all the participants, not just the ones who experienced a positive result.

Feeling the Impact of Misleading Statistics

How do misleading statistics affect your life? They can affect you in small ways or in large ways, depending on the type of statistics that cross your path and what you choose to do with the information that you're given. The most important way that statistics (good or bad) affect you is in your everyday decision making.

Think about the examples discussed throughout this chapter and how they could affect your decision making. You probably won't stay up at night wondering whether the remaining 14% of those surveyed actually microwave their leftovers. But you may run into other situations involving statistics that can affect you greatly, and you need to be ready and able to sort it all out. Here are some examples:

✔ Someone may try to tell you that four out of five people surveyed agree that taxes should be raised, so you should too! Will you feel the pressure, or will you try to find out more information first? (Were you one of those kids that lived on the phrase "everyone else is doing it"?)

✔ A political candidate sends you a newsletter giving campaign information based on statistics. Can you believe what he/she is telling you?

✔ If you're ever chosen to be on a jury, chances are that somewhere along the line, you'll see a lawyer use statistics as part of an argument. You have to sort through all of the information and determine whether the evidence is convincing "beyond a reasonable doubt." In other words, what's the chance that the defendant is guilty? (For more on how to interpret probabilities, see Chapters 7 and 8.)

✔ The radio news on the top of the hour says cellphones cause brain tumors. Your spouse uses his or her cellphone all the time. Should you be concerned?

✔ What about those endless drug company advertisements? Imagine the pressure doctors must feel from their patients who come in convinced by advertisements that they need to take certain medications *now*. Being informed is one thing, but feeling informed because of an ad sponsored by the maker of the product is another.

✔ If you have a medical problem, or know someone who does, you may be on the lookout for new treatments or therapies that could help. The world of medical results is full of statistics that can be very confusing.

In life, you come across everything from honest arithmetic errors to exaggerations and stretches of the truth, data fudging, *data fishing* (fishing for results), and reports that conveniently leave out information or communicate only those parts of the results that the researcher wants you to hear. While I need to stress that not all statistics are misleading and not everyone is out to get you, you do need to be vigilant. Sort out the good information from the suspicious and bad information, and you can steer clear of statistics that go wrong.

Sneaking statistics into everyday life

You make decisions every day based on statistics and statistical studies that you've heard about or seen, many times without even realizing it. Here are some examples of the types of decisions you may be making as you go through the day.

✔ "Should I wear boots today? What did the weather report say last night? Oh yeah, a 30% chance of snow."

✔ "How much water should I be drinking? I used to think eight eight-ounce glasses was the best thing to do, but now I hear that too much water could be bad for me!"

✔ "Should I go out and buy some vitamins today? Mary says they work for her, but vitamins upset my stomach." (When *is* the best time to take a vitamin, anyway?)

✔ "I'm getting a headache; maybe I should take an aspirin. Maybe I should try being out in the sun more, I've heard that helps stop migraines."

✔ "Gee, I hope Rex doesn't chew up my rugs again while I'm at work. I heard somewhere that dogs on Prozac deal better with separation anxiety. A dog on Prozac? How would

they get the dosage right? And what would I tell my friends?"

✔ "Should I just do drive through again for lunch? I've heard of something called 'bad cholesterol.' But I suppose all fast food is the same — bad for you, right?"

✔ "I wonder whether the boss is going to start cracking down on employees who send e-mail. I heard about a study that showed that people spend on average two hours a day checking and sending personal e-mails from work. No way do I spend that much time doing that!"

✔ "Not another guy weaving in and out of traffic talking on his cellphone! I wonder when they're going to outlaw cellphones! I'm sure they cause a huge number of accidents!"

Not all of the examples involve numbers, yet they all involve the subject called statistics. Statistics is really about the process of making decisions, testing theories, comparing groups or treatments, and asking questions. The number crunching goes on behind the scenes, leaving you with lasting impressions and conclusions that ultimately get embedded into your daily decisions.

Chapter 3

Tools of the Trade

· ·

In This Chapter

▶ Seeing statistics as a process, not just as numbers

▶ Getting familiar with some basic statistical jargon

· ·

In today's numbers explosion, the buzzword is data, as in, "Do you have any data to support your claim?" "What data do you have on this?" "The data supported the original hypothesis that . . . ," "Statistical data show that . . . ," and "The data bear this out . . . ." But the field of statistics is not just about data. Statistics is the entire process involved in gathering evidence to answer questions about the world, in cases where that evidence happens to be numerical data.

In this chapter, you see firsthand how statistics works as a process and where the numbers play their part. You also get in on the most commonly used forms of statistical jargon, and you find out how these definitions and concepts all fit together as part of that process. So, the next time you hear someone say, "This survey had a margin of error of plus or minus 3 percentage points," you'll have a basic idea of what that means.

Statistics: More than Just Numbers

Most statisticians don't want statistics to be thought of as "just statistics." While the rest of the world views them as such, statisticians don't think of themselves as number crunchers; more often, they think of themselves as the keepers of the scientific method. (Of course, statisticians depend on experts in other fields to supply the interesting questions, because man cannot live by statistics alone.) The *scientific method* (asking questions, doing studies, collecting evidence, analyzing that evidence, and making conclusions) is something you may have come across before, but you may also be wondering what this method has to do with statistics.

All research starts with a question, such as:

- ✔ Is it possible to drink too much water?
- ✔ What's the cost of living in San Francisco?
- ✔ Who will win the next presidential election?
- ✔ Do herbs really help maintain good health?
- ✔ Will my favorite TV show get renewed for next year?

None of these questions asks anything directly about numbers. Yet each question requires the use of data and statistical processes to come up with the answer.

Suppose a researcher wants to determine who will win the next U.S. presidential election. To answer this question with confidence, the researcher has to follow several steps:

1. **Determine the group of people to be studied.**

   In this case, the researcher would use registered voters who plan to vote in the next election.

2. **Collect the data.**

   This step is a challenge, because you can't go out and ask every person in the United States whether they plan to vote, and if so, for whom they plan to vote. Beyond that, suppose someone says, "Yes, I plan to vote." Will that person really vote come Election Day? And will that same person tell you for whom he or she actually plans to vote? And what if that person changes his or her mind later on and votes for a different candidate?

3. **Organize, summarize, and analyze the data.**

   After the researcher has gone out and gotten the data that she needs, getting it organized, summarized, and analyzed helps the researcher answer her question. This is what most people recognize as the business of statistics.

4. **Take all the data summaries, the charts and graphs, and the analyses, and draw conclusions from them to try to answer the researcher's original question.**

   Of course, the researcher will not be able to have 100% confidence that her answer is correct, because not every person in the United States was asked. But she can get an answer that she can be *nearly* 100% sure is the correct answer. In fact, with a sample of about 2,500 people who are selected in a fair and *unbiased* way (so that each member of the population has an equal chance of being selected), the researcher can get accurate results, within plus or minus 2.5% (that is, if all of the steps in the research process are done correctly).

In making conclusions, the researcher has to be aware that every study has limits, and that — because there is always a chance for error — the results could be wrong. A numerical value can be reported that tells others how confident the researcher is about the results, and how accurate these results are expected to be. (See Chapter 10 for more information on margin of error.)

After the research is done and the question has been answered, the results typically lead to even more questions and even more research. For example, if men appear to favor Miss Calculation but women favor her opponent, the next questions could be, "Who goes to the polls more often on Election Day — men or women — and what factors determine whether they will vote?"

The field of statistics is really the business of using the scientific method to answer research questions about the world. Statistical methods are involved in every step of a good study, from designing the research to collecting the data to organizing and summarizing the information to doing an analysis, drawing conclusions, discussing limitations and, finally, to designing the next study in order to answer new questions that arise. Statistics is more than just a number, it's a process!

Grabbing Some Basic Statistical Jargon

Every trade has a basic set of tools, and statistics is no different. If you think about the statistical process as a series of stages that one goes through to get from a question to an answer, you may guess that at each stage, you'll find a group of tools and a set of terms (or statistical jargon) to go along with it. Now if the hair is beginning to stand up on the back of your neck, don't worry. No one is asking you to become a statistics expert and plunge into the heavy-duty stuff, and no one is asking you to become a statistics nerd and use this jargon all the time. And you don't have to carry a calculator and pocket protector in your front left pocket like statisticians do, either.

But as the world becomes more numbers-conscious, statistical terms are thrown around more in the media and in the workplace, so knowing what the language really means can give you a leg up. Also, if you're reading this book because you want to find out more about how to calculate some simple statistics, understanding some of the basic jargon is your first step. So, in this section, you get a taste of statistical jargon; I send you to the appropriate chapters elsewhere in the book to get details.

Population

For virtually any question that you may want to investigate about the world, you have to center your attention on a particular group of individuals (for example, a particular group of people, cities, animals, rock specimens, exam scores, and so on). For example:

- ✔ What do Americans think about the president's foreign policy?
- ✔ What percentage of the planted crops in Wisconsin were destroyed by deer last year?
- ✔ What's the prognosis for breast cancer patients taking a new experimental drug?
- ✔ What percentage of all toothpaste tubes get filled according to their specifications?

In each of these examples, a question is posed. And in each case, you can identify a specific group of individuals who are being studied: the American people, all planted crops in Wisconsin, all breast cancer patients, and all toothpaste tubes that are being filled, respectively. The group of individuals that you wish to study in order to answer your research question is called a *population*. Populations, however, can be hard to define. In a good study, researchers define the population very clearly, while in a bad study, the population is poorly defined.

The question about whether babies sleep better with music is a good example of how difficult defining the population can be. Exactly how would you define a baby? Under 3 months old? Under a year old? And do you want to study babies only in the United States, or do you want to study all babies worldwide? The results may be different for older and younger babies, for American versus European versus African babies, and so on.

 Many times, researchers want to study and make conclusions about a broad population, but in the end, in order to save time, money, or just because they don't know any better, they study only a narrowly defined population. That can lead to big trouble when conclusions are drawn. For example, suppose a college professor wants to study how TV ads persuade consumers to buy products. Her study is based on a group of her own students who participated in order to get five points extra credit (you know you're one of them!). This may be a convenient sample, but her results can't be generalized to any population beyond her own students, because no other population was represented in her study.

Sample

When you sample some soup, what do you do? You stir the pot, reach in with a spoon, take out a little bit of the soup, and taste it. Then you draw a conclusion about the whole pot of soup, without actually having tasted all of it. If your sample is taken in a fair way (for example, you didn't just grab all the good stuff) you will get a good idea how the soup tastes without having to eat it all. This is what's done in statistics. Researchers want to find out something about a population, but they don't have time or money to study every single individual in the population. So what do they do? They select a small number of individuals from the population, study those individuals, and use that information to draw conclusions about the whole population. This is called a *sample.*

Sounds nice and neat, right? Unfortunately it's not. Notice that I said *select* a sample. That sounds like a simple process, but in fact, it isn't. The way a sample is selected from the population can mean the difference between results that are correct and fair and results that are garbage. As an example, suppose you want to get a sample of teenagers' opinions on whether they're spending too much time on the Internet. If you send out a survey over e-mail, your results won't represent the opinions of *all teenagers,* which is your intended population. They will represent only those teenagers who have Internet access. Does this sort of statistical mismatch happen often? You bet.

One of the biggest culprits of statistical misrepresentation caused by bad sampling is surveys done on the Internet. You can find thousands of examples of surveys on the Internet that are done by having people log on to a particular Web site and give their opinions. Even if 50,000 people in the United States complete a survey on the Internet, it doesn't represent the population of all Americans. It represents only those folks who have Internet access, who logged on to that particular Web site, and who were interested enough to participate in the survey (which typically means that they have strong opinions about the topic in question).

The next time you're hit with the results of a study, find out the makeup of the sample of participants and ask yourself whether this sample represents the intended population. Be wary of any conclusions being made about a broader population than what was actually studied. (More in Chapter 16.)

Random

A *random sample* is a good thing; it gives every member of the population an equal chance of being selected, and it uses some mechanism of chance to choose them. What this really means is that people don't select themselves to participate, and no one in the population is favored over another individual in the selection process.

As an example of how the experts do it, here is the way The Gallup Organization does its random sampling process. It starts with a computerized list of all telephone exchanges in America, along with estimates of the number of residential households that have those exchanges. The computer uses a procedure called *random digit dialing* (RDD) to randomly create phone numbers from those exchanges, and then selects samples of telephone numbers from those. So what really happens is that the computer creates a list of *all possible* household phone numbers in America, and then selects a subset of numbers from that list for Gallup to call. (Note that some of these phone numbers may not yet be assigned to a household, creating some logistical issues to deal with.)

Another example of random sampling involves the manufacturing sector and the concept of quality control. Most manufacturers have strict specifications for their products being produced, and errors in the process can cost them money, time, and credibility. Many companies try to head off problems before they get too large by monitoring their processes and using statistics to make decisions as to whether the process is operating correctly or needs to be stopped. For more on quality control and statistics, see Chapter 19.

Examples of *non-random* (in other words *bad*) sampling include samples from polls for which you phone in your opinion. This is not truly a random sample because it doesn't give everyone in the population an equal opportunity to participate in the survey. (If you have to buy the newspaper or watch that TV show, and then agree to write in or call in, that gives you a big clue that the sampling process is not random.) For more on sampling and polls, see Chapter 16.

Any time you look at results of a study that were based on a sample of individuals, read the fine print, and look for the term "random sample." If you see that term, dig further into the fine print to see how the sample was actually selected and use the preceding definition to verify that the sample was, in fact, selected randomly.

Bias

Bias is a word you hear all the time, and you probably know that it means something bad. But what really constitutes bias? *Bias* is systematic favoritism that is present in the data collection process, resulting in lopsided, misleading results.

Bias can occur in any of a number of ways.

> ✔ **In the way the sample is selected:** For example, if you want to get an estimate of how much Christmas shopping people in your community plan to do this year, and you take your clipboard and head out to the mall on the day after Thanksgiving to ask customers about their shopping plans, you have bias in your sampling process. Your sample tends to favor those die-hard shoppers at that particular mall who were braving the massive crowds that day.

✔ **In the way data are collected:** Poll questions are a major source of bias. Because researchers are often looking for a particular result, the questions they ask can often reflect that expected result. For example, the issue of a tax levy to help support local schools is something every voter faces at one time or another. A poll question asking, "Don't you think it would be a great investment in our future to support the local schools?" does have a bit of bias. On the other hand, so does the question, "Aren't you tired of paying money out of your pocket to educate other people's children besides your own?" Question wording can have a huge impact on the results. See Chapter 16 for more on designing polls and surveys.

When examining polling results that are important to you or that you're particularly interested in, find out what questions were asked and exactly how the questions were worded before drawing your conclusions about the results.

Data

Data are the actual measurements that you get through your study. (Remember that "data" is plural — the singular is *datum* — so sentences that use that word always sound a little funny, but they are grammatically correct.) Most data fall into one of two groups: numerical data or categorical data (see Chapter 5 for additional information).

✔ *Numerical data* are data that have meaning as a measurement, such as a person's height, weight, IQ, or blood pressure; the number of stocks a person owns; the number of teeth a person's dog has; or anything else that can be counted. (Statisticians also refer to numerical data as *quantitative data* or *measurement data*.)

✔ *Categorical data* represent characteristics, such as a person's gender, opinion, race, or even bellybutton orientation (innie versus outie — is nothing sacred anymore?). While these characteristics can take on numerical values (such as a "1" indicating male and "2" indicating female), those numbers don't have any specific meaning. You couldn't add them together, for example. (Note that statisticians also call this *qualitative data*.)

Not all data are created equal. Finding out how the data were collected can go a long way toward determining how you weigh the results and what conclusions you draw from them.

Data set

A *data set* is the collection of all the data taken from your sample. For example, if you measured the weights of five packages, and those weights were 12 lbs, 15 lbs, 22 lbs, 68 lbs, and 3 lbs, those five numbers (12, 15, 22, 68, 3) constitute your data set. Most data sets are quite a bit larger than this one, however.

Statistic

A statistic is a number that summarizes the data collected from a sample. People use many different statistics to summarize data. For example, data can be summarized as a percentage (60% of the households sampled from the United States own more than two cars), an average (the average price of a home in this sample is . . .), a median (the median salary for the 1,000 computer scientists in this sample was . . .), or a percentile (your baby's weight is at the 90th percentile this month, based on data collected from over 10,000 babies . . .).

Not all statistics are correct or fair, of course. Just because someone gives you a statistic, nothing guarantees that the statistic is scientific or legitimate! You may have heard the saying, "Figures don't lie, but liars figure."

Statistics are based on sample data, not on population data. If you collect data from the entire population, this process is called a *census*. If you then summarize all of the census information into one number, that number is a *parameter,* not a statistic. Most of the time, researchers are trying to estimate the parameters using statistics. In the case of the U.S. Census Bureau, that agency wants to report the total number of people in the United States, so it conducts a census. However, due to logistical problems in doing such an arduous task (such as being able to contact homeless folks), the census numbers can only be called estimates in the end, and they're adjusted upward to account for those people that the census missed. The long form for the census is filled out by a random sample of households; the U.S. Census Bureau uses this information to draw conclusions about the entire population (without asking every person to fill out the long form).

Mean (average)

The *mean,* also referred to by statisticians as the *average,* is the most common statistic used to measure the center, or middle, of a numerical data set. The mean is the sum of all the numbers divided by the total number of numbers. See Chapter 5 for more on the mean.

The mean may not be a fair representation of the data, because the average is easily influenced by *outliers* (very large or very small values in the data set that are not typical).

Median

The median is another way to measure the center of a numerical data set (besides the good old standby, the average). A statistical median is much like the median of an interstate highway. On a highway, the median is the middle of the road, and an equal number of lanes lay on either side of the median. In a numerical data set, the *median* is the point at which there are an equal number of data points whose values lie above and below the median value. Thus, the median is truly the middle of the data set. See Chapter 5 for more on the median.

The next time you hear an average reported, look to see whether the median is also reported. If not, ask for it! The average and the median are two different representations of the middle of a data set and can often give two very different stories about the data.

Standard deviation

Have you heard anyone report that a certain result was found to be "2 standard deviations above the mean"? More and more, people want to report how significant their results are, and the number of standard deviations above or below average is one way to do it. But exactly what *is* a standard deviation?

The *standard deviation* is a way statisticians use to measure the amount of variability (or spread) among the numbers in a data set. As the term implies, a standard deviation is a standard (or typical) amount of deviation (or distance) from the average (or mean, as statisticians like to call it). So, the standard deviation, in very rough terms, is the average distance from the mean. See Chapter 5 for calculations and more information.

The standard deviation is also used to describe where most of the data should fall, in a relative sense, compared to the average. For example, in many cases, about 95% of the data will lie within two standard deviations of the mean. (This result is called the empirical rule. See Chapter 8 for more on this.)

The formula for standard deviation(s) is as follows:

$$s = \sqrt{\frac{\sum(x - \bar{x})^2}{n - 1}}, \text{ where}$$

n = the number of values in the data set
$\bar{x}$ = the average of all the values
x = each value in the data set

For detailed instructions on calculating the standard deviation, see Chapter 5.

The standard deviation is an important statistic, but it is often absent when statistical results are reported. Without it, you're getting only part of the story about the data. Statisticians like to tell the story about the man who had one foot in a bucket of ice water and the other foot in a bucket of boiling water. He said that, on average, he felt just great! But think about the variability in the two temperatures for each of his feet. Closer to home, the average house price, for example, tells you nothing about the range of house prices you may encounter when house-hunting. The average salary may not fully represent what's really going on in your company, if the salaries are extremely spread out.

Don't be satisfied with finding out only the average — be sure to ask for the standard deviation, as well. Without a standard deviation, you have no way of knowing how spread out the values may be. (If you're talking starting salaries, for example, this could be very important!)

Percentile

You've probably heard references to percentiles before. If you've taken any kind of standardized test, you know that when your score was reported, it was presented to you with a measure of where you stood, compared to the other people who took the test. This comparison measure was most likely reported to you in terms of a percentile. The *percentile* reported for a given score is the percentage of values in the data set that fall below that certain score. For example, if your score was reported to be at the 90th percentile, that means that 90% of the other people who took the test with you scored lower than you did (and 10% scored higher than you did). For more specifics on percentiles, see Chapter 5.

Percentiles are used in a variety of ways for comparison purposes and to determine *relative standing* (that is, how an individual data value compares to the rest of the group). Babies' weights are often reported in terms of percentiles, for example. Percentiles are also used by companies to get a handle on where they stand compared to other companies in terms of sales, profits, customer satisfaction, and so on.

Standard score

The standard score is a slick way to put results in perspective without having to provide a lot of details — something that the media loves. The *standard score* represents the number of standard deviations above or below the mean (without caring what that standard deviation or mean actually are).

As an example, suppose Bob took his statewide 10th-grade test recently, and scored 400. What does that mean? It may not mean much to you because you can't put that 400 into perspective. But knowing that Bob's standard score on the test is +2 tells you everything. It tells you that Bob's score is 2 standard deviations above the mean. (Bravo, Bob!) Now suppose Bill's standard score is –2. In this case, this is not good (for Bill), because it means Bill's score is 2 standard deviations *below* the mean.

The formula for standard score is

$$\text{Standard score} = \frac{\left(\text{original score} - \bar{x}\right)}{s}, \text{ where}$$

$\bar{x}$ is the average of all the scores
s is the standard deviation of all the scores

For the details on calculating and interpreting standard scores, see Chapter 8.

Normal distribution (or bell-shaped curve)

When numerical data are organized, they're often ordered from smallest to largest, broken into reasonably sized groups, then put into graphs and charts to examine the shape, or distribution, of the data. The most common type of data distribution is called the *bell-shaped curve,* in which most of the data are centered around the average in a big lump, and as you move farther out on either side of the mean, you find fewer and fewer data points. Figure 3-1 shows a picture of a bell-shaped curve; notice that the shape of the curve resembles the outline of an old-fashioned bell.

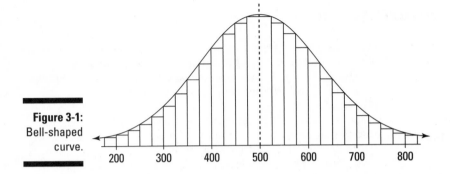

Figure 3-1:
Bell-shaped
curve.

Statisticians have another name for the bell-shaped curve when many possible values for the data exist; they call it the *normal distribution*. This distribution is used to describe data that follow a bell-shaped pattern, including what the range of values is expected to be and where an individual score stands in relation to the others. For example, if the data have a normal distribution, you can expect most of the data to lie within two standard deviations of the mean. Because every distinct population of data has a different mean and standard deviation, an infinite number of different normal distributions exist, each with its own mean and its own standard deviation to characterize it. See Chapter 8 for plenty more information on the normal distribution.

The normal distribution is also used to help measure the accuracy of many statistics, including the mean, using an important result in statistics called the *central limit theorem*. This theorem gives you the ability to measure how much your sample mean will vary, without having to take any other sample means to compare it with (thankfully!). It basically says that your sample mean has a normal distribution, no matter what the distribution of the original data looks like (as long as your sample size was large enough). See Chapter 9 for more on the central limit theorem (known by statisticians as the "crown jewel of all statistics." Should you even bother to tell them to get a life?).

If a data set has a normal distribution, and you standardize all of the data to obtain standard scores, those standard scores are called Z-values. Z-values have what is known as a standard normal distribution (or Z-distribution). The *standard normal distribution* is a special normal distribution with a mean equal to 0 and a standard deviation equal to 1. The standard normal distribution is useful for examining the data and determining statistics like percentiles, or the percentage of the data falling between two values. So if researchers determine that the data have a normal distribution, they will usually first standardize the data (by converting each data point into a Z-value), and then use the standard normal distribution to explore and discuss the data in more detail.

Experiments

An *experiment* is a study that imposes a certain amount of control on the study's subjects and their environment (for example, restricting their diets, giving them certain dosage levels of a drug or placebo, or asking them to stay awake for a prescribed period of time). The purpose of most experiments is to pinpoint a cause-and-effect relationship between two variables (such as alcohol consumption and impaired vision). Here are some of the questions that experiments try to answer:

- ✔ Does taking zinc help reduce the duration of a cold? Some studies show that it does.

- ✔ Does the shape and position of your pillow affect how well you sleep at night? The Emory Spine Center in Atlanta says, "Yes."

- ✔ Does shoe heel height affect foot comfort? A study done at UCLA says up to one inch heels are better than flat soles.

In this section, you find more information about how experimental studies are (or should be) conducted. And Chapter 17 is entirely dedicated to the subject. For now, just concentrate on the basic lingo relating to experiments.

Treatment group versus control group

Most experiments try to determine whether some type of treatment (or important factor) has some sort of effect on an outcome. For example, does zinc help to reduce the length of a cold? Subjects who are chosen to partici- pate in the experiment are typically divided into two groups, a treatment group and a control group. The *treatment group* consists of those who receive the treatment that supposedly has an effect on the outcome (in this case, zinc). The *control group* consists of those who do not receive the treatment, or those who receive a standard, well-known treatment whose results will be compared with this new treatment (such as vitamin C, in the case of the zinc study).

Placebo

A *placebo* is a fake treatment, such as a sugar pill. It is often given to the members of the control group, so that they will not know whether they are taking the treatment (for example, zinc) or receiving no treatment at all. Placebos are given to the control group in order to control for a phenomena called the *placebo effect,* in which patients who receive any sort of perceived treatment by taking a pill (even though it's a sugar pill) report some sort of result, be it positive ("Yes, I feel better already") or negative ("Wow, I am starting to feel a bit dizzy"), due to a psychological effect. Without a placebo, the researchers could not be certain that the results were due to the actual effect of the treatment, because some (or all) of the observed effect could have been due to the placebo effect.

Blind and double-blind

A *blind experiment* is one in which the subjects who are participating in the study are not aware of whether they're in the treatment group or the control group. In the zinc example, a placebo would be used that would look like the zinc pill, and patients would not be told which type of pill they were taking. A blind experiment attempts to eliminate any bias in what the study subjects might report.

A *double-blind experiment* controls for potential bias on the part of both the patients and the researchers. Neither the patients nor the researchers collecting the data know which subjects received the treatment and which ones didn't. A double-blind study is best, because even though researchers may claim to be unbiased, they often have a special interest in the results — otherwise they wouldn't be doing the study!

Surveys (polls)

A *survey* (more commonly known as a *poll*) is a measurement tool that is most often used to gather people's opinions along with some relevant demographic information. Because so many policymakers, marketers, and others want to "get at the pulse of the American public" and find out what the average American is thinking and feeling, many people now feel that they cannot escape the barrage of requests to take part in surveys and polls. In fact, you've probably received many requests to participate in surveys, and you may even have become numb to them, simply throwing away surveys received in the mail, or saying "no" when you're asked to participate in a telephone survey.

If done properly, a survey can really be informative. People use surveys to find out what TV programs Americans (and others) like how consumers feel about Internet shopping and whether the United States should have a nuclear defense system. Surveys are used by companies to assess the level of satisfaction their customers feel, to find out what products their customers want, and to determine who is buying their products. TV stations use surveys to get instant reactions to news stories and events, and movie producers use them to determine how to end their movies.

If I had to choose one word to describe the general state of surveys in the media today, I'd have to use the word *quantity,* rather than *quality.* In other words, you'll find no shortage of bad surveys. You can ask a few basic questions to determine whether a survey has been conducted properly; these issues are covered in detail in Chapter 16.

Estimation

One of the biggest uses of statistics is to guesstimate something (the statistical term is *estimation*), as in the following examples:

- What's the average household income in America?
- What percentage of households tuned in to the Academy Awards this year?
- What's the average life expectancy of a baby born today?
- How effective is this new drug?
- How clean is the air today, compared to ten years ago?

All of these questions require some sort of numerical estimate to answer the question, yet the business of coming up with a fair and accurate estimate can be quite involved. The following sections cover major elements in that process. For more information on making and interpreting estimates, see Chapter 11.

Margin of error

You've probably heard someone report, "This survey had a margin of error of plus or minus 3 percentage points." What does this mean? All surveys are based on information collected from a sample of individuals, not the entire population. A certain amount of error is bound to occur — not in the sense of calculation error (although there may be some of that, too) but in the sense of *sampling error,* or error that's bound to happen simply because the researchers aren't asking everyone. The *margin of error* is supposed to measure the maximum amount by which the sample results are expected to differ from those of the actual population. Because the results of most survey questions can be reported in terms of percentages, the margin of error most often appears as a percentage, as well.

How do you interpret a margin of error? Suppose you know that 51% of those sampled say that they plan to vote for Miss Calculation in the upcoming election. Now, projecting these results to the whole voting population, you would have to add and subtract the margin of error and give a range of possible results in order to have sufficient confidence that you're bridging the gap between your sample and the population. So, in this case (supposing a margin of error of plus or minus 3 percentage points) you would be pretty confident that between 48% and 54% of the population will vote for Miss Calculation in the election, based on the sample results. In this case, Miss Calculation may get slightly more or slightly less than the majority of votes and could either win or lose the election. This has become a familiar situation in recent years, where the media want to report results on Election Night, but based on survey results, the election is "too close to call." For more on the margin of error, see Chapter 10.

The margin of error measures accuracy; it does not measure the amount of bias that may be present. Results that look numerically scientific and precise don't mean anything if they were collected in a biased way.

Confidence interval

When you combine your estimate with the margin of error, you come up with a *confidence interval*. For example, suppose the average time it takes you to drive to work each day is 35 minutes, with a margin of error of plus or minus 5 minutes. You estimate that the average time to work would be anywhere from 30 to 40 minutes. This estimate is a confidence interval. It takes into account the fact that sample results will vary and gives an indication of how much variation to expect. For more on confidence interval basics, see Chapter 11.

Some confidence intervals are wider than others (and wide isn't good, because it equals less accuracy). Several factors influence the width of a confidence interval, such as sample size, the amount of variability in the population being studied, and how confident you want to be in your results. (Most researchers are happy with a 95% level of confidence in their results.) For more on factors that influence confidence intervals, see Chapter 12.

Many different types of confidence intervals are done in scientific research, including confidence intervals for means, proportions, the difference of two means or proportions, or paired differences. For specifics on the most common hypothesis tests, see Chapter 13.

Probability versus odds

A *probability* is a measurement of the likelihood of an event happening. In other words, a probability is the chance that something will happen. For example, if the chance of rain tomorrow is 30%, it's less likely to rain than not rain tomorrow, but the chance of rain is still 3 out of 10. (Given those chances, will you bring your umbrella with you tomorrow?) A chance of rain of 30% also means that over many, many days with the same conditions as those predicted for tomorrow, it rained 30% of the time.

Probabilities are calculated in many different ways:

✔ Math is used to grind out the numbers (for example, figuring your chances of winning the lottery or determining the hierarchy of hands in poker).

✔ Data are collected, and the probabilities are estimated based on the history of the data (for example, to predict the weather).

✔ Complex math and computer models are used to try to predict future behavior and occurrence of natural phenomena (for example, hurricanes and earthquakes).

The laws of probability often go against your intuition and your own beliefs about what you think can happen (that's why casinos stay in business). See Chapter 6 for more on probability.

Odds and probability are slightly different. The best way to describe this difference is by looking at an example. Suppose the probability that a certain race horse is going win the race is 1 out of 10. That means his probability of winning is 1 in 10 or 1 ÷ 10 or 0.10. A probability reflects the chances of winning. Now what are this horse's odds of winning? They are 9 to 1. That's because odds are actually a ratio of the chances of losing to the chances of winning. This horse has a 9 in 10 chance of losing and a 1 in 10 chance of winning. Take $\frac{9}{10}$ over $\frac{1}{10}$ and the 10s cancel, leaving you with $\frac{9}{1}$, which in odds lingo is stated as "9 to 1." For more on gambling, see Chapter 7.

The law of averages

You've probably heard people mention the law of averages before. Perhaps it was the local baseball reporter lamenting that his team, who defied the odds by winning 50 games and losing only 12 in the first 3 months of the season, were now starting to lose, giving in to the law of averages. Or maybe the context was gambling ("The law of averages is bound to catch up with me — I'm on too hot of a winning streak!"). What is the law of averages, exactly, and are people using this term properly?

The *law of averages* is a rule of probability. It says that, in the long term, results will average out to their expected value, but in the short term, no one knows what will happen. For example, casinos set up all of their games so that the chances of the house winning are slightly in their favor. That means that in the long term, as long as people keep playing, the casinos are going to come out ahead, on average. Of course there will be some winners, that's what keeps people playing, knowing they could be among them. But in the long term, the losers outweigh the winners (not to mention the fact that many times people who win big just put their money back into the games again and end up losing). For more on the law of averages, see Chapter 7.

Hypothesis testing

Hypothesis test is a term you probably haven't run across in your everyday dealings with numbers and statistics. But I guarantee that hypothesis tests have been a big part of your life and your workplace, simply because of the major role they play in industry, medicine, agriculture, government, and a host of other areas. Any time you hear someone talking about their results showing a "statistically significant difference," you're encountering the results of a hypothesis test. Basically, a *hypothesis test* is a statistical procedure in which data are collected and measured against a claim about a population.

For example, if a pizza delivery chain claims to deliver pizzas within 30 minutes of placing the order, you could test whether this claim is true by collecting a random sample of delivery times over a certain period of time and looking at the average delivery time for that sample.

Because your decision is based on a sample and not the entire population, a hypothesis test can sometimes lead you to the wrong conclusion. However, statistics are all you have, and if done properly, they can get as close to the truth as is humanly possible without actually knowing the truth. For more on the basics of hypothesis testing, see Chapter 14.

A variety of hypothesis tests are done in scientific research, including t-tests, paired t-tests, and tests of proportions or means for one or more populations. For specifics on the most common hypothesis tests, see Chapter 15.

P-value

Hypothesis tests are used to confirm or deny a claim that is made about a population. This claim that's on trial, in essence, is called the *null hypothesis.* The evidence in the trial is your data and the statistics that go along with it. All hypothesis tests ultimately use a *p*-value to weigh the strength of the evidence (what the data are telling you about the population). The *p*-value is a number between 0 and 1 that reflects the strength of the data that are being used to evaluate the null hypothesis. If the *p*-value is small, you have strong evidence against the null hypothesis. A large *p*-value indicates weak evidence against the null hypothesis. For example, if a pizza chain claims to deliver pizzas in less than 30 minutes (this is the null hypothesis), and your random sample of 100 delivery times has an average of 40 minutes for the delivery time (which is more than 2 standard deviations above what the average delivery time is supposed to be) the *p*-value for this test would be small, and you would say you have strong evidence against the pizza chain's claim.

Statistically significant

Whenever data are collected to perform a hypothesis test, the researcher is usually typically looking for a significant result. Usually, this means that the researcher has found something out of the ordinary. (Research that simply confirms something that was already well known doesn't make headlines, unfortunately.) A *statistically significant* result is one that would have had a very small probability of happening just by chance. The *p*-value reflects that probability.

For example, if a drug is found to be more effective at treating breast cancer than the current treatment is, researchers say that the new drug shows a statistically significant improvement in the survival rate of patients with breast cancer (or something to that effect). That means that based on their data, the difference in the results from patients on the new drug compared to those using the old treatment is so big that it would be hard to say it was just a coincidence.

Sometimes, a sample doesn't represent the population (just by chance) and this results in a wrong conclusion. For example, a positive effect that's experienced by a sample of people who took the new treatment may have just been a fluke. (Assume for the moment that you know that the data were not fabricated, fudged, or exaggerated.) The beauty of medical research is that as soon as someone gives a press release saying that he or she found something significant, the rush is on to try to replicate the results, and if the results can't be replicated, this probably means that the original results were wrong, for some reason. Unfortunately, a press release announcing a "major breakthrough" tends to get a lot of play in the media, but follow-up studies refuting those results often don't show up on the front page.

One statistically significant result shouldn't lead to quick decisions on anyone's part. In science, what counts is not a single remarkable study, but a body of evidence that is built up over time, along with a variety of well-designed follow-up studies. Take any major breakthroughs you hear about with a grain of salt and wait until the follow-up work has been done before using the information from a single study to make important decisions in your life.

Correlation and causation

Of all of the misunderstood statistical issues, the most problematic is the misuse of the concepts of correlation and causation.

Correlation means that two numerical variables have some sort of linear relationship. For example, the number of times crickets chirp per second is related to temperature; when it's cold outside, they chirp less frequently, and when it's warm outside they chirp more frequently. (This actually happens to be true!) Another example of correlation has to do with police staffing. The number of crimes (per capita) has often been found to be related to the number of police officers in a given area. When more police officers patrol an area, crime tends to be lower, and when fewer police officers are present, crime tends to be higher. However, seemingly unrelated events have also been found to be correlated. One such example is the consumption of ice cream (pints per person) and the number of murders in certain areas. Now maybe having more police officers deters crime, but does having people eat less ice cream deter crime? What's the difference? The difference is that with correlation, a link or relationship is found to exist between two variables, x and y. With *causation,* one makes that leap and says "a change in x will cause a change in y to happen." Too many times in research, in the media, or in the public consumption of statistical results, that leap is made when it shouldn't be. When can it be done? When a well-designed experiment is conducted that eliminates any other factors that could have been related to the outcomes. For more on correlation and causation, see Chapter 18.

Part II
Number-Crunching
Basics

"GET READY, I THINK THEY'RE STARTING TO DRIFT."

In this part . . .

Number crunching: It's a dirty job, but somebody has to do it. Why not let it be you? Even if you aren't a numbers person and calculations aren't your thing, the step-by-step approach in this part may be just what you need to boost your confidence in doing and really understanding statistics.

In this part, you get down to the basics of number crunching, from making and interpreting charts and graphs to cranking out and understanding means, medians, standard deviations, and more. You also develop important skills for critiquing someone else's statistical information and getting at the real truth behind the data.

Chapter 4

Getting the Picture: Charts and Graphs

Someone once said that a picture is worth a thousand words. In statistics, a picture may be worth a thousand data points — as long as that picture is done correctly, of course. Data displays, such as charts and graphs, appear often in everyday life. These displays show everything from election results, broken down by every conceivable characteristic, to how the stock market has fared over the past few years. Today's society is a fast-food, fast-information society; everyone wants to know the bottom line and be spared the details. The main use of statistics is to boil down information into summary form, and data displays are a natural way to do that. But do data displays give you the whole picture of what's happening with the data? That depends on the quality of the data display and its intended purpose. Pictures can be misleading (sometimes intentionally and sometimes by accident), and not every data display that you see will be correct. This chapter helps you gain a better understanding of the use of charts and graphs in the media and the workplace and shows you how to read and make sense of these data displays. In this chapter, I also give you some tips for evaluating data displays and for spotting those (oh so many!) misleading displays.

Getting Graphic with Statistics

The main purpose of a data display is to make a certain point, make the point clearly and effectively, and make the point correctly. A chart or graph, for example, is used to give impact to a specific characteristic of the data, highlight changes over time, compare and contrast opinions or demographic data, or show links between pieces of information. Data displays "break down" a statistical story that the author wants to relay about a data set, so that the reader can quickly see the issue at a glance and come to some conclusion. For this reason, data displays are powerful: Used properly, they can be informative and effective; used improperly, they can be misleading and destructive.

Data displays can impact your life in large and small ways — responding to these critically and understanding what they say and don't say helps you become a savvy consumer of statistical data. You want to become familiar with the various types of data displays that you're likely to come across and to explore how these displays are used in the media and the workplace.

Researchers and journalists use different ways to display each of the two major types of data: categorical data, which represent qualities or characteristics (such as gender or political party) and numerical data, which represent measured quantities (such as height or income).

The most common types of data displays for categorical data are as follows:

- ✔ Pie charts (see the "Getting a Piece of the Pie Chart" section)
- ✔ Bar graphs (see the "Raising the Bar on Bar Graphs" section)
- ✔ Tables (see the "Putting Statistics on the Table" section)
- ✔ Time charts, also called *line graphs* (see the "Keeping Pace with Time Charts" section)

For numerical data, tables are commonly used to display the data. In addition, histograms should be commonly used to display numerical data (but often aren't), so I include them in the "Picturing Data with a Histogram" section.

In this chapter, I present examples of each type of data display, some thoughts on interpretation, and tips for critically evaluating each type.

Getting a Piece of the Pie Chart

The pie chart is one of the most commonly used data displays because it's easy to read and can quickly make a point. You most likely have seen them before — they seem so simple. Can anything go wrong with an innocent pie chart? The answer is yes.

A pie chart takes categorical data and breaks them down by group, showing the percentage of individuals that fall into each group. Because a pie chart takes on the shape of a circle, or pie, the "slices" that represent each group can easily be compared and contrasted to one another. Because each individual in the group falls into one and only one category, the sum of all the slices of the pie should be 100% or close to it (subject to a bit of round-off error).

Tallying personal expenses

When you spend your money, what do you spend it on? What are the top three expenses that you have? According to the U.S. Bureau of Labor Statistics, the top three sources of consumer expenditures in 1994 were housing (32%), transportation (19%), and food (14%). Figure 4-1 shows these results in a pie chart. (Notice that the "other" category is a bit large in this chart. But in this case determining which of the other items should be included as a single pie chart slice would be difficult, because so many different types of expenditures for different people are possible.)

How did the U.S. government get this information? From something called the Consumer Expenditure Survey. Many federal agencies are charged with collecting data (often using surveys) and disseminating the results through written reports. (The U.S. government is a good source of information about many aspects of everyday life in the United States.)

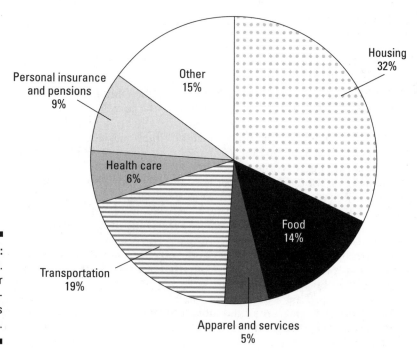

Figure 4-1:
Total U.S. consumer expenditures for 1994.

Sizing up the lottery

State lotteries bring in a great deal of revenue, and they also return a large portion of the money received, with some of the revenues going to prizes and some being allocated to state programs, such as education. Where does the money come from? Figure 4-2 shows a pie chart showing types of games and percentage of revenue they generate for the Ohio lottery.

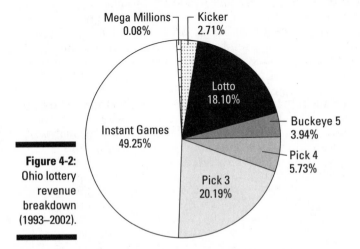

Figure 4-2: Ohio lottery revenue breakdown (1993–2002).

You can see by this pie chart that most of the Ohio lottery sales revenues (49.25%) come from the instant (scratch-off) games. The rest of the revenues come from various lottery-type games in which players choose a set of numbers and win if a certain number of their numbers match those chosen by the lottery. Why do the instant games account for such a large portion of the lottery sales? One possible reason is that the payouts for instant games are frequent, even though they're not very big. Also, you get instant feedback with the scratch-off games; with the lottery games you have to wait until a drawing occurs before you know whether you're a winner. On the other hand, maybe people just enjoy the satisfaction of scratching off those boxes!

Notice that this pie chart doesn't tell you *how much* money came in, only *what percentage* of the money came from each type of game. In other words, you know how the pie is divided up, but you don't really know how big the pie is to begin with. This is something you may want to know as a consumer of this information. About half of the money (49.25%) came from instant scratch off games; does this revenue represent a million dollars, two million dollars, ten million dollars, or more? The pie chart in Figure 4-2 doesn't tell you that information, and you can't determine it on your own without being given the total amount of revenue dollars. I was, however, able to find this information on another chart provided by the Ohio lottery: The total revenue for 2002 from Ohio lottery sales was reported as "1,983.1 million dollars" — which you also know as 1.983 billion dollars. Because 49.25% of sales came from instant games, this represents a sales revenue of $976,676,750 over a 10-year period. That's a lot of scratching!

Pie charts often show the breakdown of the portion or percentage of the total that falls in each group or category. But they often do not show you the total number in each group, in terms of original units (number of dollars, number of people, and so on). This approach results in a loss of information, may not necessarily present the whole story behind the data, and leaves you wondering what the total amount is that's being divided up. You can always go from amounts to percents, but you can't go from percentages back to the original amounts without knowing a total. With survey results, this lack of information can be a real problem; oftentimes, pie charts show the percentage of people who answered the question in a certain way, but they don't tell you how many people responded to the survey — a critical piece of information needed to assess the accuracy of the results. (See Chapter 10 for more on accuracy and margin of error for surveys.)

Always look for the total number of individuals when given any data display. If it's not directly available, ask for it!

The Florida lottery uses a pie chart to report where your money goes when you purchase one of its lottery tickets (see Figure 4-3). You can see that half of the Florida lottery revenues (50 cents of every dollar spent) goes to prizes, and 38 cents of every dollar goes to education. This pie chart does break down the way each dollar of revenue is spent, but you probably also want to know *how many* dollars are spent playing the Florida lottery. Florida lottery ticket sales for 2001 actually totaled $2,360.6 million (or $2.36 billion), which amounts to $147.70 per capita (that is, per person), as shown in Table 4-1.

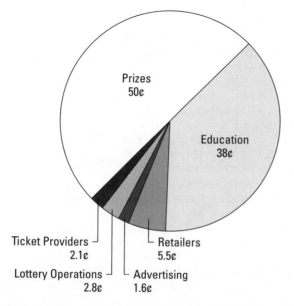

Figure 4-3: Florida lottery expenditures (fiscal year 2001–2002).

Table 4-1   Top Ten Lotteries (2001)

Rank	Lottery	Population (Millions)	Ticket Sales (Millions)	Prizes (Millions)	Net Income (Millions)	Prize (% of Revenues)	Sales per Capita ($)
1	New York	18.976	4,178	2,274	1,447	54.4%	220.16
2	Mass.	6.349	3,923	2,774	865	70.7%	617.85
3	California	33.872	2,896	1,492	1,048	51.5%	85.49
4	Texas	20.852	2,826	1,639	865	58.0%	135.50
5	Florida	15.982	2,361	1,180	862	50.0%	147.70
6	Georgia	8.186	2,194	1,142	692	52.0%	267.98
7	Ohio	11.353	1,920	1,113	637	58.0%	169.11
8	New Jersey	8.414	1,807	991	695	54.8%	214.72
9	Pennsylvania	12.281	1,780	996	627	55.9%	144.93
10	Michigan	9.938	1,615	874	586	54.1%	162.49

Interestingly, the Web site for the Michigan lottery reports the amount, in dollars, that the lottery gives to education each year, but not the percentage of the total lottery revenue that goes to education. For example, the 2001 amount reportedly given to education by the Michigan lottery was $587 million. Because you know from Table 4-1 that the total lottery sales revenue for Michigan was $1,615 million (aka $1.6 billion), you can calculate the percentage of revenue that was given to education in this state. In Michigan, about 36% ($587 million ÷ $1,615 million × 100%) of the lottery sales revenue was given to education.

Pie charts are easy to use to compare the sizes of slices within a single pie itself, but they can also be used to compare one entire pie to another. For example, the New York lottery reports its expenditures using a pie chart (see Figure 4-4).

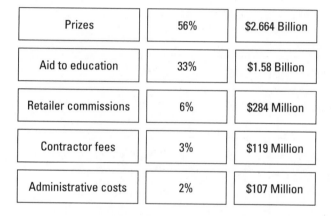

Prizes	56%	$2.664 Billion
Aid to education	33%	$1.58 Billion
Retailer commissions	6%	$284 Million
Contractor fees	3%	$119 Million
Administrative costs	2%	$107 Million

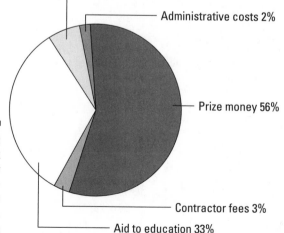

Figure 4-4: New York lottery expenditures (2001–2002).

Comparing Figures 4-3 and 4-4, you can see that for the New York lottery, 56% of the money goes to prizes (slightly more than the percentage for Florida), and 33% goes to education (slightly less than the percentage for Florida). Included with each of the New York lottery's pie charts is a table showing the actual dollar amounts, allowing you to see more of the whole story. (However, the New York lottery makes you add the total up for yourself; it's well over $4.5 billion dollars.)

The state of New York also wants you to realize how much money it's putting toward education, in terms of a piece of the school-revenue pie (which is a very smart move, politically speaking). Figure 4-5 shows that whereas 4% of New York school revenue in 2001–2002 came from federal aid, 5% came from the New York lottery. Again, this pie chart also comes with a table showing the actual dollar amounts. (In actuality, the only amount that's needed is the grand total dollar amount, because from the grand total and the percents in the pie chart, you can generate the numbers in the table. Not having to do that extra work is nice, however.)

Local revenue	45%	$15.168 Billion
Other state aid to education	42%	$14.148 Billion
New York Lottery	5%	$1.58 Billion
Federal aid	4%	$1.48 Billion
Other sources	4%	$1.43 Billion

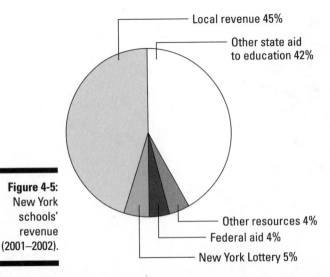

Figure 4-5: New York schools' revenue (2001–2002).

Local revenue 45%

Other state aid to education 42%

Other resources 4%

Federal aid 4%

New York Lottery 5%

Slicing up your tax dollars

The Internal Revenue Service (IRS) wants you to know where your tax money goes, and if you tell them how much money you paid in taxes last year, they will show you how your tax dollars were sliced up. Figure 4-6 shows an example of the results that you get from the IRS if you tell them you paid $10,000 in taxes last year.

This data display is creative but a bit different (dare I say "odd," at the risk of being audited?). First, this chart appears to be more of a pizza chart than a pie chart. But you have to ask what the pizza is doing there, if it's not being sliced up to show you where your money is going. The percentages are in the table next to the pizza, so they're available. This chart would have more visual impact if the IRS had shown the actual "slices" of pizza that correspond to the percentages in the table. What's nice about this display, however, is that it shows the amount of money as well as the percentages that are spent in each area. (By the way, no matter what the total amount of tax dollars is, the percentages showing where the money is allocated don't change; only the dollar amounts do.)

Federal Spending

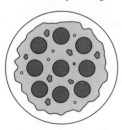

Your Money Is Spent On	Your Share	Percentages
National defense	$1,700.00	17%
Medicaid	$700.00	7%
Medicare	$1,200.00	12%
Unemployment, disability and other income	$1,400.00	14%
Social Security	$2,300.00	23%
Interest payments	$1,100.00	11%
Other expenses	$1,600.00	16%
Total paid	**$10,000.00**	**100%**

Figure 4-6:
How the tax dollar is sliced up (2002).

Examining Figure 4-6, you can see that the biggest slice of your tax dollars go to Social Security (23%), and the second biggest slice goes to national defense (17%). It seems strange, though, that the IRS breaks down certain categories as low as single digits (for example 7% going to Medicaid), but the third highest slice of the pie (or in this case, pizza) actually shows up as "other expenses" (16%).

Ideally, a pie chart doesn't have too many slices because a large number of slices distracts the reader from the big issues that the pie chart is trying to relay. However, if lumping all of those remaining categories into a category called "other" results in a category that's one of the largest ones in the whole pie chart, readers are left wondering what's included in that slice of pie.

Perhaps you're wondering what those "other expenses" are in the IRS chart. If you probe further on the IRS Web site, the IRS tells you that "other expenses" means "federal employment retirement benefits, payments to farmers, and other activities." This doesn't provide a great deal of additional information, but maybe that's all you really want to know. In fairness to the IRS, I'm sure the details are all spelled out in some neatly filed government report.

Predicting population trends

The U.S. Census Bureau provides many data displays in its reports about the U.S. population. Figure 4-7 shows two pie charts comparing the racial breakdown of the United States in 1995 (actual figures) with the projected racial breakdown in 2050, if current trends continue. You can see that in 1995, about 73.6% of the U.S. population was White, while Blacks made up the second highest group at 12.0%, closely followed by those of Hispanic origin, who comprised 10.2% of the population. (Note that although Hispanics are typically white or black, they are shown here as a separate category, independent of racial background.) The Census Bureau projects that Whites will be a declining share of the total U.S. population in the future, whereas the Hispanic share of the population will grow faster than that of non-Hispanic Blacks. This point is made well using the two pie charts, as opposed to tables simply showing the percentages.

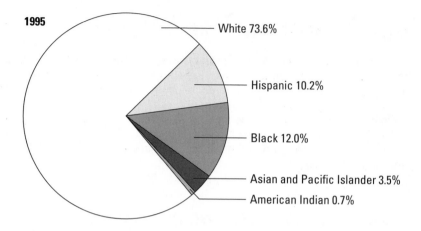

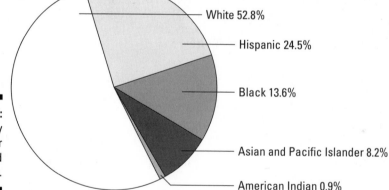

Figure 4-7: Ethnicity trends for the United States.

Evaluating a pie chart

To taste test a pie chart for statistical correctness:

✔ Check to be sure the percentages add up to 100% or close to it (any round-off error should be very small).

✔ Beware of slices of the pie called "other" that are larger than many of the other slices.

✔ Look for a reported total number of units, so that you can determine how big the pie was before being divided up into the slices that you're looking at.

Raising the Bar on Bar Graphs

A bar graph, or bar chart, is perhaps the most common data display used by the media. Like a pie chart, a bar graph breaks categorical data down by group, showing how many are in each group. A bar graph, however, represents those groups by using bars of different lengths, rather than as pie slices of varying sizes. And whereas a pie chart most often reports the amount in each group as percentages, a bar graph uses either the number of individuals in each group or the percentage of the total in each group. In a bar graph, the length of each bar indicates the number or percent in each group.

Tracking transportation expenses

How much of their income do people spend on transportation? It depends on how much money they make. The Bureau of Transportation Statistics (did you know such a department existed?) conducted a study on transportation in the United States in 1994, and many of their findings are presented as bar graphs like the one shown in Figure 4-8.

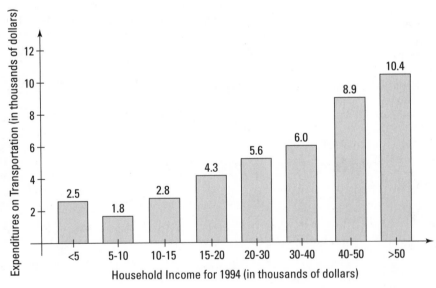

Figure 4-8:
Transportation expenses by household income for 1994.

This particular bar graph shows how much money is spent on transportation for people in varying household-income groups. It appears that as household income increases, the total expenditures on transportation also increase. This probably makes sense, because the more money people have, the more they have available to spend. But would the bar graph change if you looked at transportation expenditures not in terms of total dollar amounts, but as the percentage of household income? The households in the first group make less than $5,000 a year and have to spend $2,500 on transportation per year. (Notice that the table reads "2.5," but because the units are in thousands of dollars, the 2.5 translates into $2,500.) This $2,500 represents 50% of the annual income of those who make $5,000 per year; it's an even higher percentage of the total income for those who make less than $5,000 per year. The households earning $30,000–$40,000 per year pay $6,000 per year on transportation, which is between 15% and 20% of their household income. So, although the people making more money spend more dollars on transportation, they don't spend more as a percentage of their total income. Depending on how you look at expenditures, the bar graph can tell two somewhat different stories.

This bar graph has another peculiarity. The categories for household income as shown aren't equivalent. For example, each of the first four bars represents household incomes in intervals of $5,000, but the next three groups increase by $10,000 each, and the last group contains every household making more than $50,000 per year, which is a large percentage of households, even in 1994. Bar graphs with different category ranges, such as the one shown in Figure 4-8, make comparison between groups more difficult.

Highlighting mothers in the workforce

Bar graphs are often used to compare two groups by breaking down the categories for each group and showing them as side-by-side bars. One example of this is shown in Figure 4-9, which asks the question, "Has the percentage of mothers in the workforce changed over time?" The answer is yes. Figure 4-9 shows that the overall percentage of mothers in the workforce climbed from 47% to 72% between 1975 and 1998. Taking the age of the child into account, fewer mothers work while their children are younger and not in school yet, but the difference from 1975 to 1998 is still about 25% in each case, as shown by the side-by-side bars.

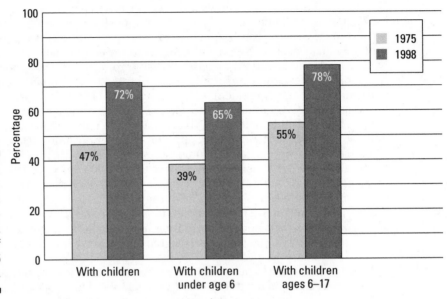

Playing with the Ohio lottery

The Ohio lottery shows its sales and expenditures for 2002 using a bar graph (see Figure 4-10). This bar graph takes some additional work behind the scenes to make it understandable. The first issue with this bar graph is that the bars don't represent similar types of entities. The first bar represents sales (a form of revenue), and the other bars represent expenditures. This bar graph could be made more clear if the first bar weren't included; for example, the total sales could be listed as a footnote. Also, the expenditures could be represented in a pie chart, as is done by some of the other state lotteries (refer to Figures 4-3 and 4-4). The next issue is that the sum of all of the expenditures ($2,013.2 million — in other words, $2.0132 billion) is greater than the sales ($1.9831 billion), so some additional revenue is not being shown in this bar graph (either that, or the Ohio lottery is about to go out of business!). Looking deeper into more of the information provided on the Ohio lottery Web site, I found out that besides sales, additional revenue was reported to be $124.1 million, earned through "interest and other revenues." That brings the total revenues to $1.9831 billion + $124.1 million = $2.1072 billion in 2002, leaving a profit of $2.1072 billion – $2.0132 billion = $.094 billion, or $94,000,000.

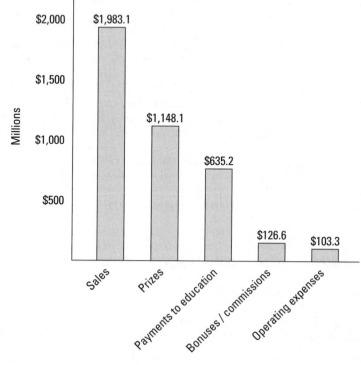

Figure 4-10:
Ohio lottery
sales and
expen-
ditures
for 2002.

Notice that throughout the Ohio lottery example, the units have been in terms of millions, so you see funny looking numbers like $1,983.1 million, which really should be written as $1.9831 billion. Why does the Ohio lottery use units in millions? Maybe to make it look like it isn't bringing in as much money as it really is. The numbers do look tamed down a bit in Figure 4-10. Don't think the lotteries fail to miss these subtleties: They are masters of subtlety. (Think about this: In order to figure out profits, you have to do the calculations yourself, and you need to go to two different places in order to get the numbers needed for those calculations. The lotteries don't make it easy!)

Don't assume that the information being presented in a data display represents everything you need to know; be prepared to dig deeper if you need to fill in any missing information (for which charts and graphs in the media are notorious!). It usually doesn't take too long to find what you're looking for (or to at least discount the information you're being presented, if what you find shows bias or inaccuracy).

Bar graphs allow a great deal of poetic license to whomever designs them. That's because the person designing the bar graph determines what scale he or she wants to use, and that means that the information can be presented in a misleading way. By using a smaller scale (for example, having each half inch of the height of a bar represent 10 units versus 50 units) you can stretch the truth, make differences look more dramatic, or exaggerate values. By using a larger scale (for example having each half inch of a bar represent 50 units versus 10 units) you can downplay differences, make results look less dramatic than they actually are, or even make small differences appear to be non-existent. (See Chapter 2 for examples of this.)

Note that in a pie chart, however, the scale can't be changed to over-emphasize (or downplay) the results. No matter how you slice up a pie chart, you're always slicing up a circle, and the proportion of the total pie belonging to any given slice won't change, even if you make the pie bigger or smaller.

Evaluating a bar graph

To raise the statistical bar on bar graphs, check out these items:

- ✔ Bars that divide up values of a numerical variable (such as income) should be equal in width for fair comparison.

- ✔ Be aware of the scale of the bar graph (the units in which heights of the bars are represented) and determine whether it's an appropriate representation of the information.

- ✔ Don't assume the information being presented in the bar graph represents everything you need to know; be prepared to dig deeper if you need to.

Putting Statistics on the Table

A *table* is a data display that presents summary information from a data set in a row-and-column format. Some tables are clear and easy to read; others leave something to be desired. Although a pie chart or a bar graph is usually intended to make one or two points at most, a table can make several points at once (which can be good or bad, depending on the effect this has on the reader).

Statistical information is compiled by researchers not only for their own reports, but also so that others can use the information to do their own research and answer their own questions. Tables are often used in these situations.

Examining birth statistics

The Colorado Department of Public Health and Environment compiles tables on birth statistics for Colorado residents. Table 4-2 shows the number of live births by the sex of the child and the plurality status (single births versus the births of twins, triplets, and so on) for selected years from 1975–2000. Some questions that can be answered with this table are: What's the birth rate of males compared to females in Colorado? And is the rate of plural births changing? From this table, you can see that over a 25-year period, the percentage of female births remained steady at just under 49%, while the percentage of male births remained steady at just over 51%. (You may wonder why these percentages aren't closer to 50% each. This is a question for demographers — scientists who study human population trends — and biologists, not statisticians.) You can also see that the rate of plural births (as opposed to single births) seems to have changed over the years. It appears that the percentage of plural births is increasing, but which column do you look at: the number of plural births or the percentage of plural births? Does it matter? Yes, it does!

Looking at percents versus totals

How do you draw conclusions about trends in plural births over time by using the statistics presented in this table? If you look only at the number of plural births for 1975 compared to 2000, they increase from 763 to 1,982. Someone may try to say that this represents a 160% increase, or about 1.6 times as many plural births in 25 years ([1,982 – 763] ÷ 763). More plural births occurred in the year 2000 than in 1975, but more single births also occurred over this same time period. Because of this, the only accurate way to compare these statistics is to calculate the percentage of single versus plural births and compare these percentages. Looking at Table 4-2, you can see that the percentage of plural births in 1975 was 1.9%, while in 2000 the percentage of plural births was 3.0%. You can conclude that the percentage of plural births did increase over time, even after taking the increased number of births into account. However, the increase is not 160%; it's closer to 58%: ([3.0 – 1.9]) ÷ 1.9) × 100%.

Beware of conclusions that are drawn from a data display that compares the *number* of individuals, as opposed to the *percentage* of individuals. Percentages represent a relative comparison of quantities (often over a period of time); this is usually an accurate way of comparing quantities, especially when the total number of items or events also changes over time. By looking at the percent change, you take into account the fact that the total number has also changed. (Of course, if you're truly interested in examining how the *number* of each item changes, you should look at the numbers and not the percentages.)

Table 4-2 Colorado Live Births by Sex and Plurality Status

Year	Total Number of Births	Number of Female Births	% Female Births	Number of Male Births	% Male Births	Number of Single Births	% Single Births	Number of Plural Births	% Plural Births
1975	40,148	19,447	48.4	20,701	51.6	39,385	98.1	763	1.9
1980	49,716	24,282	48.8	25,434	51.2	48,771	98.1	945	1.9
1985	55,115	26,925	48.9	28,190	51.1	53,949	97.9	1,166	2.1
1990	53,491	26,097	48.8	27,394	51.2	52,245	97.7	1,246	2.3
1995	54,310	26,431	48.7	27,879	51.3	52,669	97.0	1,641	3.0
2000	65,429	31,953	48.8	33,476	51.2	63,447	97.0	1,982	3.0

Table 4-3 shows a breakdown of the number of live births in Colorado by the age of the mother for selected years from 1975–2000. The numerical variable age is broken down into categories that are of the same width (5 years) and are not overlapping. This makes for a fair and equitable comparison of age groups. However, the table gives only numbers of births in each case, so you can't look at the table and get a sense of any trends that may be developing over time in terms of the age of the mother. This problem can be solved by including the percents in parentheses along with the total number in each category, so that the reader can easily make a comparison. Another way to display the information is to include a pie chart for each year, showing the percent of the total births that were born to women in each of the eight non-overlapping age groups.

Table 4-3		Colorado Live Births by Mother's Age							
Year	Total Number of Births*	Age of Mother (Years)							
		10–14	15–19	20–24	25–29	30–34	35–39	40–44	45–49
1975	40,148	88	6,627	14,533	12,565	4,885	1,211	222	16
1980	49,716	57	6,530	16,642	16,081	8,349	1,842	198	12
1985	55,115	90	5,634	16,242	18,065	11,231	3,464	370	13
1990	53,491	91	5,975	13,118	16,352	12,444	4,772	717	15
1995	54,310	134	6,462	12,935	14,286	13,186	6,184	1,071	38
2000	65,429	117	7,546	15,865	17,408	15,275	7,546	1,545	93

*Note: The sums of the births may not add up to the total number of births, due to unknown or unusually high age (50 or over) of the mother.

Note that because the totals are reported in this table, you can do the work of finding the percentages on your own, if you wanted to. (Had the table presented only percents, without any totals, you would have had an easier time comparing percents. But you would have been limited in the conclusions you could have drawn, because you would not have known the total numbers.) Just to save you time, I calculated those percents for you, for the combined group of mothers aged 40–49. Table 4-4 shows those calculations. From this table, you can see that a trend in mother's age appears to be emerging. More women are having babies in their 40s than before, and the percentage is steadily increasing.

The footnote to Table 4-3 (paraphrased from the note originally written by the Colorado Department of Public Health and Environment) indicates that any mothers who were aged 50 or older were not included in this data set. Recent studies suggest that a growing (albeit still small) percentage of women are having babies in their early 50s, so this data set may have to be augmented in time to include that age group.

Table 4-4		Percent of Colorado Live Births to Mothers Aged 40-49	
Year	*Total Births*	*Number of Births to Mothers Aged 40–49 Years*	*% of Births to Mothers Aged 40–49 Years*
1975	40,148	238	0.59%
1980	49,716	210	0.42%
1985	55,115	383	0.69%
1990	53,491	732	1.4%
1995	54,310	1,109	2.0%
2000	65,429	1,638	2.5%

Putting percents into perspective

Don't be fooled into thinking that just because certain percentages are small, they aren't meaningful and/or comparable. All of the percentages in Table 4-4 are small (equal to or less than 2.5%) but the percentage for the year 2000 (2.5%) is over four times the percentage for 1975 (0.59%), and that represents a very large increase, relatively speaking. Similarly, don't assume that when a large percent increase is reported, the situation involves a large number of people. Suppose someone announces that the rate of a particular disease quadrupled over the past few years. That doesn't mean a large percentage of people are affected, it means only that the percentage is four times as large as it used to be. The percentage of people affected by the disease in question may have been extremely small to begin with. An increase is still an increase, but in some situations, reporting the percentage alone can be misleading; the prevalence of the disease needs to be put into perspective in terms of the total number of people affected.

A percentage is a relative measure. However, look for the total number, as well, to keep the actual amounts in proper perspective.

Keeping an eye on the units

Sometimes, tables can be a bit confusing if you're not watching carefully. For example, the IRS reports "Tax Stats at a Glance" on its Web site, and some of those statistics (reported exactly as the IRS did) are shown in Table 4-5.

Table 4-5	Statistics on Individual Income Tax Returns
Number of Returns (FY2001)	129,783,221
Gross Collections (FY2001 in millions of dollars)	1,178,210
Top 1% AGI break (TY1999)	$293,415
Top 10% AGI break (TY1999)	$87,682
Bottom 10% AGI break (TY1999)	$4,718
Median adjusted gross income (AGI, TY2000)	$27,355
Percent claiming standard deductions (TY2000)	66.2%
Percent claiming itemized deductions (TY2000)	32.9%
Percent using paid preparers (TY2000)	53.4%
Percent e-filed (TY2001) thru 5/3/2002	38.3%
Number of returns with AGI > $1million (TY2000)	241,068
Number of individual refunds (TY2000 in millions)	93.0
Individual refund amount (TY2000 in billions of dollars)	167.6

Two features of this table are noticeable right away. First, the IRS reports statistics for many different reporting years on the same table, for example FY2001 (which means fiscal year 2001, or the 12-month period from July 1, 2000 through June 30, 2001), TY1999 (which means tax year, or calendar year, 1999), and TY 2000. Note that the tax year and the fiscal year overlap, and that the tax returns for TY 2000, for example, are due to the IRS in April, 2001 (which is in FY 2001). How's that for confusion? Also notice that you can't compare the median adjusted gross income (median AGI) in this table to the top 1% or top 10% AGI in the same table, because these figures are listed for different years (TY 2000 and TY 1999, respectively).

Second, the way the IRS reports the dollar units can lead to confusion. For example, the gross collections for individual tax returns for FY 2001 (reported to be in millions of dollars) is listed as 1,178,210. This means that $1,178,210 *million* dollars were collected from individual tax returns. Now you typically don't display dollar amounts that way. To put this into proper perspective, $1,178,210 million is actually $1,178,210,000,000, which is $1.178 trillion. No wonder the IRS didn't report these revenues that way, it's too big of a number to even fathom!

When looking at a table, be sure you understand the units that are being expressed and watch for changes in units (such as the year) throughout the table.

Table 4-5 is intended to show several points; here are a few of them. In the tax year 2000, the ratio of people who claimed the standard deduction relative to people who itemized their deductions, is about 2 to 1. (A pie chart would show this very nicely.) About half of the people who filed for the tax year 2000 used tax preparers, and the percentage who e-filed for the tax year 2001 was about 38% (a percentage the IRS probably would like to see increase over time). The average refund for the tax year 2000 was $1,802.15 (the total amount of refund dollars in billions, divided by the total number of refunds, in millions). (Now, isn't it more impressive to report the total amount of refund dollars, as opposed to the average refund amount?) Also notice that the IRS didn't report the total number of tax returns filed for the tax year 2000 or the percentage of tax filers who received a refund for that tax year. Knowing this percent would be more useful than knowing the total number of refunds in a given tax year. After all, not everyone got a refund — plenty of people had to pay or paid exactly the right amount.

Tables are designed to make certain points more prominent and to make other points less noticeable. Sometimes, the de-emphasized information — or even the missing information — is the most telling!

Evaluating a table

To find out whether a table is sturdy enough, statistically:

- ✔ Know the difference between percentages and total numbers and how these two statistics are used to interpret the results. Percentages are often the most sensible statistic to use for comparing different results.

- ✔ With numerical data, be sure that the groups in the table don't overlap and that the groups are divided evenly for an equitable comparison.

- ✔ Look closely at the units and how they're presented in the table.

- ✔ Look at the way the information is presented. Often, tables are designed to downplay certain points while highlighting only the points that the researchers or reporters want you to notice.

Keeping Pace with Time Charts

A *time chart* is a data display whose main point is to examine trends over time. Another name for a time chart is a *line graph*. Typically a time chart will have some unit of time on the horizontal axis (such as year, day, month, and so on) and some measured quantity on the vertical axis (such as average household income, birth rate, total sales, percentage of people in favor of the president, and so on). At each time period, the amount is represented by a dot, and the dots are connected to form the time chart.

Analyzing wage trends

In 1999, the U.S. Bureau of Labor Statistics put out a report on work trends in the United States, and what the outlook was for the future. Its report includes many time charts, including the two shown in Figure 4-11 and Figure 4-12. Figure 4-11 shows the trend over time in the average hourly wage for production workers from 1947 to 1998. (Because of inflation, it wouldn't make sense to simply show the actual hourly wages over this time period. You want to know the information in terms of "real wages" — in other words, something comparable over time. Here, the bureau shows everything in terms of 1998 dollars for equitable comparison over time.) You can see from Figure 4-11 that wages for production workers increased from 1947 until the early 1970s, declined during the 1970s, and basically stayed in the same range until the late 1990s, when a small surge began.

Figure 4-12 makes the point that the gap in wages between educated and non-educated workers has widened between 1979 and 1997.

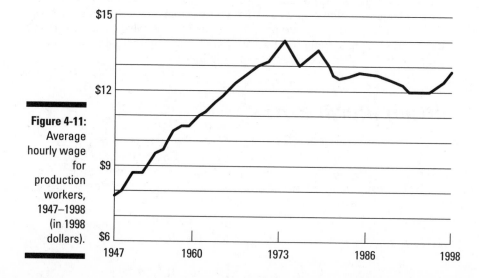

Figure 4-11: Average hourly wage for production workers, 1947–1998 (in 1998 dollars).

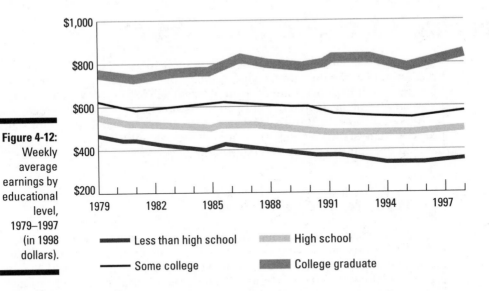

Figure 4-12:
Weekly
average
earnings by
educational
level,
1979–1997
(in 1998
dollars).

Statistics tell you facts — in other words, they tell you what's occurring. But they don't explain why events are occurring as they are. The report put out by the Bureau of Labor Statistics not only presents data showing wage trends, but also goes beyond data presentation to discuss some of the reasons *why* the average production wage stagnated from the late '70s to the mid '90s and *why* the earnings gap for more-educated versus less-educated workers is growing. Answering the "why" question is much more complex than answering the "what" question. Although the Bureau of Labor Statistics certainly has other statistics to back up its assessment of why the trends are doing what they're doing, not everyone who puts out statistics does.

Many folks try to take a simple data display and use it not only to show what's happening, but also to try to explain why things are happening as they are. Without sufficient data, these people may be making false conclusions. If you suspect someone is going too far with his or her conclusions, you need to question whether these conclusions are justified.

Charting plural births

When showing birth statistics for Colorado residents, a time chart can be used to examine the trend in the rate of plural births over time. Such a time chart would look like the one in Figure 4-13. You can see that the percentage of plural births appears to be increasing over time, especially if you simply compare 1975 to 2000.

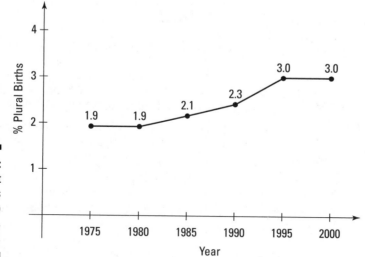

Figure 4-13:
Percent
plural births
for Colorado
residents,
1975–2000.

As with a bar graph, differences that are represented using a time chart can be played up or down by changing the scale on the vertical axis, so be sure to take the scale into account when interpreting the results of a time chart.

For the Colorado plural births example, the increase in plural births over time can be made to look more dramatic by changing the increments on the vertical axis from 1% to 0.2%, which would stretch out the graph vertically. Likewise, the trend could be made to look almost nonexistent by changing the increments on the vertical axis from 1% to 5%. The most reasonable increment is something between 0.2% and 5%. I chose 1% increments for my version of this graph.

Another factor to consider with time charts is the time axis, or how close together or far apart the data points in the chart are. Data that are close together in time but have very different quantities will make the time chart appear jagged. This is typical of time charts that represent volatile data, such as stock market prices over the course of a day, as you may have seen on television business news programs. Other time charts may show more sweeping, long-term changes, such as those in Figure 4-13, which shows only the percentage of plural births every five years, rather than every single year. Again, this depends on how the person designing the chart wants you to look at the information. You need to really think about how the chart was set up and, potentially, ask additional questions in order to clear up any confusion.

You can also run into a problem in which a time chart presents the information in an unfair way, such as charting the *number* of crimes over time, rather than the crime *rate* (number of crimes per capita). Make sure that you understand what statistics are being presented in the chart, and then examine them for fairness and appropriateness (see Chapter 2 for more on this).

Evaluating a time chart

To see whether a time chart is on pace, statistically:

- ✔ Examine the scale on the vertical (quantity) axis as well as the horizontal (timeline) axis; results can be made to look more or less dramatic than they actually are simply by changing the scale.

- ✔ Take into account the units used in the chart and be sure they're appropriate for comparison over time (for example, are dollar amounts being adjusted for inflation?).

- ✔ Beware of people trying to explain why a trend is occurring without using additional statistics to back up their claims. A time chart generally shows what is happening. Why something is happening is another story!

Picturing Data with a Histogram

Numerical data in their raw, unorganized form are hard to absorb. For example, look at Table 4-6, which shows the 2000 population estimates for each of the 50 states (and the District of Columbia), put together by the U.S. Census Bureau. Stare at the table for 30 seconds or so. After you've done that that, go ahead and try to answer these questions quickly:

- ✔ Which states have the largest/smallest populations?

- ✔ How many people reside in most of the states? Give a rough range of values.

- ✔ How much variability exists between state populations? (Are the states very similar, or very different, in terms of their total population?)

Table 4-6	Population Estimates by State (2000 Census)
State	*Census 2000 Population*
Alabama	4,447,100
Alaska	626,932
Arizona	5,130,632
Arkansas	2,673,400
California	33,871,648
Colorado	4,301,261
Connecticut	3,405,565

State	Census 2000 Population
Delaware	783,600
District of Columbia	572,059
Florida	15,982,378
Georgia	8,186,453
Hawaii	1,211,537
Idaho	1,293,953
Illinois	12,419,293
Indiana	6,080,485
Iowa	2,926,324
Kansas	2,688,418
Kentucky	4,041,769
Louisiana	4,468,976
Maine	1,274,923
Maryland	5,296,486
Massachusetts	6,349,097
Michigan	9,938,444
Minnesota	4,919,479
Mississippi	2,844,658
Missouri	5,595,211
Montana	902,195
Nebraska	1,711,263
Nevada	1,998,257
New Hampshire	1,235,786
New Jersey	8,414,350
New Mexico	1,819,046
New York	18,976,457
North Carolina	8,049,313
North Dakota	642,200

(continued)

Table 4-6 *(continued)*	
State	*Census 2000 Population*
Ohio	11,353,140
Oklahoma	3,450,654
Oregon	3,421,399
Pennsylvania	12,281,054
Rhode Island	1,048,319
South Carolina	4,012,012
South Dakota	754,844
Tennessee	5,689,283
Texas	20,851,820
Utah	2,233,169
Vermont	608,827
Virginia	7,078,515
Washington	5,894,121
West Virginia	1,808,344
Wisconsin	5,363,675
Wyoming	493,782
U.S. TOTAL	**281,421,906**

Without some way of organizing these data, you have difficulty answering these questions. Although most of the media favors the use of tables to organize numerical data, statisticians favor the histogram as their data display of choice for these kind of data. What is a histogram, you ask?

A *histogram* is basically a bar graph that applies to numerical data. Because the data are numerical, the categories are ordered from smallest to largest (as opposed to categorical data, such as gender, which has no inherent order to it). And because you want to be sure each number falls into exactly one group, the bars on a histogram touch each other but don't overlap. Each bar is marked on the x-axis (or horizontal axis) by the value representing its midpoint. For example, suppose a histogram showing length of time until failure of a car part (in hours) has two adjacent bars marked with midpoints of 1,000 hours and 2,000 hours, and each bar has a width of 500 hours. This means the first bar represents car parts that lasted anywhere from 500 to 1,500 hours, and the second bar represents car parts that lasted anywhere from 1,500 to 2,500 hours. (Numbers on the border can go on either side, as long as you're consistent for all the borderline values.)

The height of each bar of a histogram represents either the number of individuals in each group (also known as the *frequency* of each group) or the percentage of individuals in each group (also known as the *relative frequency* of each group). For example, if 50% of the car parts lasted between 500 and 1,500 hours, the first bar in the preceding example would have a relative frequency of 50%, and the height of that bar would be reflective of that.

You can see a histogram of the state population data in Figure 4-14. You can easily answer most of the questions at the beginning of this section by looking quickly at the histogram. And in my opinion, in many situations, a histogram provides a more interesting organizational summary of a data set than a table does.

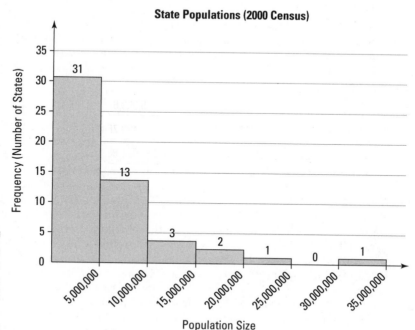

Figure 4-14:
State
population
sizes (2000
Census).

A majority of the states and the District of Columbia (31 out of 51, or 60.8%) have fewer than 5 million people. Another 25.5% have populations of between 5 and 10 million. This means that 86.3% of the states haves fewer than 10 million people each. Each of the remaining seven states has very large populations, making the histogram look lopsided and trailing off to the right (this is called *skewed to the right*). Except for those few very large states, the populations of the states aren't as variable as you may think. The histogram doesn't tell you which state is which, of course, but a quick sorting of the original data can tell you which states are largest and smallest. The five most populous states are California, Texas, New York, Florida, and Illinois (which is closely followed by Pennsylvania). The smallest state is Wyoming with about 494,000 people.

If questions come up while you're looking at a data display, try to get access to the original data set. Researchers should be able to provide you with their data if you ask for them.

Analyzing mothers' ages

In one birth statistics example (refer to Table 4-3), the age of the mother is shown for various years from 1975 to 2000. For any year on the table, the age variable is divided into groups, and you're given the number of mothers in each group. Because you're given the total numbers, you can make a histogram of mothers' ages showing either the frequencies or the relative frequencies, whichever is the most appropriate in terms of the point you want to make.

Suppose you want to compare the ages of mothers in 1975 and 2000. You can make two histograms, one for each year, and compare the results. Figure 4-15 shows two such histograms for 1975 (top) and 2000 (bottom). Notice that the relative frequencies (or percentages) are shown on the vertical axis, and the age groups for the mothers are shown on the horizontal axis.

A histogram can summarize the features of numerical data quite well. One of the features that a histogram can show you is the so-called *shape* of the data (in other words, how the data are distributed among the groups). Are the data distributed evenly, in a uniform way? Are the data *symmetric*, meaning that the left-hand side of the histogram is a mirror image of the right-hand side of the histogram? Does the histogram have a *U-shape*, with lots of data on extreme ends and not much in the middle? Does the histogram of the data have a *bell-shape*, meaning that it looks like a mound in the middle with tails trailing off in either direction as you move away from the center? Or is the histogram *skewed*, meaning that it looks like a lopsided mound with one long tail either going off to the right (indicating the data are *skewed right*) or going off to the left (indicating the data are *skewed left*)?

Mothers' ages in Figure 4-15 for years 1975 and 2000 appear to be mostly mound-shaped, although the data for 1975 are slightly more skewed to the right, indicating that as the women got older, fewer of them had babies, relative to the situation in the year 2000. Another way of saying this is that in the year 2000, a higher proportion of older women were having babies compared to 1975.

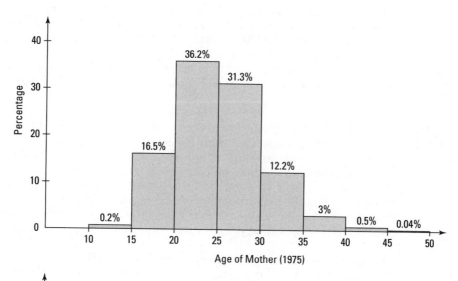

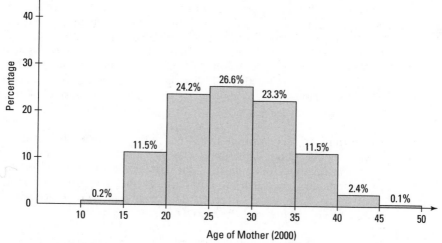

Figure 4-15: Colorado live births, by age of mother for 1975 and 2000.

You can also get a sense of how much variability exists in the data by looking at a histogram. If a histogram is quite flat with the bars close to the same height, you may think this indicates less variability because the heights of the bars are similar. In fact, the opposite is true. That's because you have an equal number in each bar, but the bars themselves represent different ranges of values, so the entire data set is actually quite spread out. Now if the histogram has a big lump in the middle with tails on the sides, this indicates that more data are in the middle bars than the outer bars, so the data are actually closer together. Comparing 1975 mothers' ages to 2000 mothers' ages, you see more variability in 2000 than in 1975. This, again, indicates changing times; more of today's women are waiting to have children, compared to 1975, when most women had their children by age 30, and the length of time they're waiting varies. (Chapter 5 shows you ways to measure variability in a data set.)

Variability in a histogram should not be confused with variability in a time chart (see the "Keeping Pace with Time Charts" section). If values change over time, they're shown on a time chart as highs and lows, and many changes from high to low (over time) indicate lots of variability. So, a flat line on a time chart indicates no change and no variability in the values across time. However, when the heights of bars of a histogram appear to be flat (uniform), this shows the opposite — the values are spread out uniformly over many groups, indicating a great deal of variability in the data at one point in time.

A histogram can also give you some idea of where the center of the data lies. The center of a data set is measured in different ways (see Chapter 5 for a discussion of these measures). One way to eyeball the center on a histogram is to think of the histogram as a picture of people sitting on a teeter-totter and the center as the point where the fulcrum has to be in order to balance the weight on each side. Refer to Figure 4-15, which shows the ages of Colorado mothers in 1975 and 2000, and note that the mid-point appears to be around 25 years for the 1975 histogram and around 27.5 years for the 2000 histogram. This suggests that in the year 2000, Colorado women were having children at older ages, on average, than they did in 1975.

Histograms aren't as commonly found in the media as they should be. The reason for this is not clear, and tables are much more commonly used to show breakdowns for numerical data. However, a histogram can be informative, especially when used to compare one group or time period to another. At any rate, if you want to look at data graphically, you can always take data from a table and convert them to a data display.

Watch for histograms that use unusual scales to mislead readers. As with bar graphs, you can exaggerate differences by using a smaller scale on the vertical axis of a histogram, and you can play down differences by using a larger scale.

Readers can be mislead by a histogram in ways that aren't possible with a bar graph. Remember that a histogram deals with numerical data, not categorical data. This means that you need to determine how you want the numerical data to be broken down into groups to display on the horizontal axis. How you determine those groupings can make the graph look very different.

Crawling with a baby

How much ground does an 8-month-old crawling baby cover? Figure 4-16 shows two histograms that represent the same data set (distances my baby crawled during a six-hour testing period). In each case, distances were rounded to the nearest foot. In the top portion of the figure, the measurements are broken into 5-foot increments, and the data seem to be distributed in a uniform way. In other words, the number of times he crawled each distance (0–5 feet, 5–10 feet, and 10–15 feet) was approximately the same. The data don't look very interesting. But in the bottom portion of the figure, I break the distances into smaller, one-foot increments, and the histogram looks different and a lot more interesting.

In this histogram, you can see two distinct groupings of distances, indicating that my baby tended to either crawl a shorter distance (around 5 feet) or a longer distance (around 10 feet) to get where he wanted to go. This makes sense because at the time I collected the data, my baby's toy basket was about 5 feet from the starting point, and the newspaper pile (another favorite toy) was 10 feet away. The second histogram is a much better representation of the data: how far my baby crawled in the given setting.

So how much ground did he cover in those six hours? Using the bottom portion of Figure 4-16, you can find the total crawling distance because the bars are in one-foot increments. Multiply the height of each bar times the distance, and then sum them all up. My baby's total crawling distance in this six-hour period was a whopping 398 feet, or 132.7 yards — more than the length of one football field!

Note that I could have broken down the distances into even smaller increments, but that would only make the histogram look cluttered and busy and wouldn't have given you any additional information. A happy medium exists in terms of the number of groupings used and the range of values they represent. Each histogram is slightly different, but somewhere between 6 and 12 groupings is a generally a good number of bars for a histogram. If the histogram has too few bars, the data don't show anything; if it has too many, the data are too disjointed and patterns get lost.

Be sure to take the scale into account on both the horizontal and vertical axes when examining the results presented in a histogram. The same data can be made to look different, depending on how they're grouped (for example, into few versus many groups) and depending on the scale of the vertical axis, which can make the bars appear taller or shorter than you'd otherwise expect.

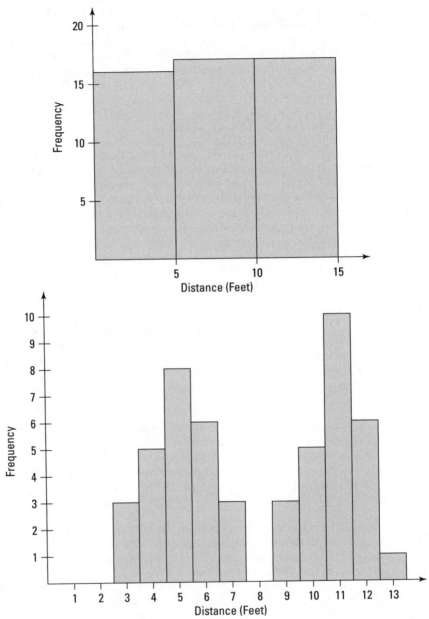

Interpreting a histogram

You can use a histogram to tell you three main features of numerical data:

- ✔ How the data are distributed (symmetric, skewed right, skewed left, bell-shaped, and so on)
- ✔ The amount of variability in the data
- ✔ Where the center of the data is (approximately)

Evaluating a histogram

To picture the statistical quality of a histogram:

- ✔ Examine the scale used for the vertical (frequency or relative frequency) axis and beware of results that appear exaggerated or played down through the use of inappropriate scales.
- ✔ Check out the units on the vertical axis to see whether the histogram reports frequencies (numbers) or relative frequencies (percentages), and then take this into account when evaluating the information.
- ✔ Look at the scale used for the groupings of the numerical variable (on the horizontal axis). If the range for each group is very small, the data may look overly volatile. If the ranges are very large, the data may appear to be smoother than they really are.

Chapter 5

Means, Medians, and More

A statistic is a number that summarizes some characteristic about a set of data. Of the hundreds of statistics that exist, a few of them are used so often that they commonly appear in the workplace and in other facets of everyday life. In this chapter, you find out which statistics are used most often, how these statistics are used, what they mean, and how they're misused.

Every data set has a story, and if used properly, statistics do a good job of telling that story. Statistics that are improperly used can tell a different story, or only part of the story, so knowing how to make good decisions about the information you're given is very important. In this chapter, you see some of the most common summary statistics. You find out more about what these summary statistics say and what they don't say about the data, which can be grouped as either numerical or categorical.

Summing Up Data with Statistics

Statistics are used to summarize some of the most basic information in a data set. Summarizing information has several different purposes. Picture your boss coming to you and asking, "What's our client base like these days and who's buying our products?" How would you like to answer that question — with a long, detailed, and complicated stream of numbers and statistics that are sure to glaze over her eyes? Probably not. You want clean, clear, and concise numbers that sum up the client base for her, so that she can see how brilliant you are, and then send you off to collect even more data to see how she can include more people in the client base. (That's what you get for being efficient.) So, statistics are often used to provide people with information that is easy to understand and that answers their questions (if answering their questions is possible).

Summarizing statistics has other purposes, as well. After all of the data have been collected from a survey or some other kind of study, the next step is for the researcher to try to make sense out of the data. Typically, the first step researchers take is to run some basic statistics on the data to get a rough idea about what's happening in the data. Later in the process, researchers can do more analyses to formulate or test claims made about the population, estimate certain characteristics about the population, look for links between items they measured, and so on.

Another big part of research is reporting the results, not only to your peers, but to the media and to the general public. While a researcher's peers may be waiting and expecting to hear about all the complex analyses that were done on a data set, the general public is neither ready for nor interested in that. What does the public want? Basic information. So, statistics that make your point clearly and concisely are commonly used to relay information to the media and to the public.

Many times, statistics are used to give a quick and dirty summary of a situation that's actually pretty complicated. In such a situation, less is not more, and sometimes the real story behind the data can get lost in the shuffle. While you have to accept that getting sound bytes of information is a fact of life these days, be sure the group putting out the data isn't watering it down at the same time. Think about which statistics are reported, what these statistics really mean, and what information is missing. This chapter focuses on these questions.

Summarizing Categorical Data

Categorical data capture qualities or characteristics about the individual, such as a person's eye color, gender, political party, or opinion on some issue (using categories such as agree, disagree, or no opinion). Categorical data tend to fall into groups or categories pretty naturally. "Political party," for example, typically has four groups: Democrat, Republican, Independent, and other. Categorical data often come from survey data, but they can also be collected in experiments. For example, in an experimental test of a new medical treatment, researchers may use three categories to assess the outcome of the experiment: Did the patient get better, worse, or stay the same while undergoing the treatment?

Categorical data are often summarized by reporting the percentage of individuals falling into each category. For example, pollsters may report the percentage of Republicans, Democrats, Independents, and others who took part in a survey. To calculate the percentage of individuals in a certain category, find the number of individuals in that category, divide by the total number of people in the study, and then multiply by 100%. For example, if a survey of 2,000 teenagers included 1,200 females and 800 males, the resulting percentages would be (1,200 ÷ 2,000) × 100% = 60% female and (800 ÷ 2,000) × 100% = 40% male.

You can further break down categorical data by creating something called crosstabs. *Crosstabs* (also called *two-way tables*) are tables with rows and columns. They summarize the information from two categorical variables at once, such as gender and political party, so you can see (or easily calculate) the percentage of individuals in each combination of categories. For example, if you had data about the gender and political party of your respondents, you would be able to look at the percentage of Republican females, Republican males, Democratic females, Democratic males, and so on. In this example, the total number of possible combinations in your table would be 2 × 4 = 8, or the total number of gender categories times the total number of party affiliation categories.

The U.S. government calculates and summarizes loads of categorical data using crosstabs. The U.S. Census Bureau doesn't just count the population; it also collects and summarizes data from a subset of all Americans (those who fill out the long form) on various demographic characteristics, such as gender and age. Typical age and gender data, reported by the U.S. Census Bureau for a survey conducted in 2001, are shown in Table 5-1. (Normally, age would be considered a numerical variable, but the way the U.S. government reports it, age is broken down into categories, making it a categorical variable. See the following section for more on numerical data.)

Table 5-1   U.S. Population, Broken Down by Age and Gender (2001)

Age	Total	%	# Males	% Males	# Females	% Females
Under 5 years	19,369,341	6.80	9,905,282	7.08	9,464,059	6.53
5 to 9 years	20,184,052	7.09	10,336,616	7.39	9,847,436	6.79
10 to 14 years	20,881,442	7.33	10,696,244	7.65	10,185,198	7.03
15 to 19 years	20,267,154	7.12	10,423,173	7.46	9,843,981	6.79
20 to 24 years	19,681,213	6.91	10,061,983	7.20	9,619,230	6.63
25 to 29 years	18,926,104	6.65	9,592,895	6.86	9,333,209	6.44
30 to 34 years	20,681,202	7.26	10,420,677	7.45	10,260,525	7.08
35 to 39 years	22,243,146	7.81	11,104,822	7.94	11,138,324	7.68
40 to 44 years	22,775,521	8.00	11,298,089	8.08	11,477,432	7.92
45 to 49 years	20,768,983	7.29	10,224,864	7.31	10,544,119	7.27

(continued)

Table 5-1 *(continued)*

Age	Total	%	# Males	% Males	# Females	% Females
50 to 54 years	18,419,209	6.47	9,011,221	6.45	9,407,988	6.49
55 to 59 years	14,190,116	4.98	6,865,439	4.91	7,324,677	5.05
60 to 64 years	11,118,462	3.90	5,288,527	3.78	5,829,935	4.02
65 to 69 years	9,532,702	3.35	4,409,658	3.15	5,123,044	3.53
70 to 74 years	8,780,521	3.08	3,887,793	2.78	4,892,728	3.37
75 to 79 years	7,424,947	2.61	3,057,402	2.19	4,367,545	3.01
80 to 84 years	5,149,013	1.81	1,929,315	1.38	3,219,698	2.22
85 to 89 years	2,887,943	1.01	926,654	0.66	1,961,289	1.35
90 to 94 years	1,175,545	0.41	303,927	0.22	871,618	0.60
95 to 99 years	291,844	0.10	58,667	0.04	233,177	0.16
100 years and over	48,427	0.02	9,860	0.01	38,567	0.03
Total, all ages	284,796,887	100	139,813,108	100	144,983,779	100

You can examine many different facets of the population by looking at and working with different numbers from Table 5-1. Looking at gender, notice that women slightly outnumber men, because the population in 2001 was 51% female (divide total number of females by total population size and multiply by 100%) and 49% male (divide total number of males by total population size and multiply by 100%). You can also look at age: The percentage of the entire population that is age 5 and under was 6.8%; the largest group belongs to the 40–44 year olds, who made up 8% of the population. Next, you can explore a possible relationship between gender and age by comparing various parts of the table. You can compare, for example, the percentage of females to males in the 80-and-over age group. Because these data are reported in five-year increments, you have to do a little math in order to get your answer, though. The percentage of the population that's female and aged 80 and above is 2.22% + 1.35% + 0.6% + 0.16% + 0.03% = 4.36%. The percentage of males aged 80 and over is 1.38% + 0.66% + 0.22% + 0.04% + 0.01% = 2.31%. This shows that the 80-and-over age group contains almost twice as many women as men. These data seem to confirm the notion that women tend to live longer than men.

If you're given the number of individuals in each group, you can always calculate your own percents. But if you're only given percentages without the total number in the group, you can never retrieve the original number of individuals in each group. For example, you could hear that 80% of the people surveyed prefer Cheesy cheese crackers over Crummy cheese crackers. But how many were surveyed? It could be only 10 people, for all you know, because 8 out of 10 is 80%, just as 800 out of 1,000 is 80%. These two fractions (8 out of 10 and 800 out of 1,000) have different meanings for statisticians, because in the first case, the statistic is based on very little data, and in the second case, it's based on a lot of data. (See Chapter 10 for more information on data accuracy and margin of error.)

After you have the crosstabs that show the breakdown of two categorical variables, you can conduct statistical tests to determine whether a significant relationship or link between the two variables exists. (See Chapter 18 for more information on these statistical tests.)

Summarizing Numerical Data

With *numerical data,* measurable characteristics such as height, weight, IQ, age, or income are represented by numbers. Because the data have numerical meaning, you can summarize them in more ways than is possible with categorical data. Certain characteristics of a numerical data set can be described using statistics, such as where the center is, how spread out the data are, and where certain milestones are. These kinds of summaries occur often in the media, so knowing what these summary statistics say and don't say about the data helps you better understand the research that's presented to you in your everyday life.

Getting centered

The most common way to summarize a numerical data set is to describe where the center is. One way of thinking about what the center of a data set means is to ask, "What's a typical value?" Or, "Where is the middle of the data?" The center of a data set can actually be measured in different ways, and the method chosen can greatly influence the conclusions people make about the data.

Averaging out NBA salaries

NBA players make a lot of money, right? Compared to most people, they certainly do. But how much do they make, and is it really as much as you think it is? The answer depends on how you choose to summarize the information. You often hear about players like Shaquille O'Neal, who made $21.4 million in the 2001–2002 season. Is that what the typical NBA player makes? No. Shaquille O'Neal was the highest paid NBA player of that season.

So how much does the typical NBA player make? One way to answer this is to look at the *average* salary. The average is probably the most commonly used statistic of all time. It is one way to determine where the "center" of the data is.

Here is what you need to do to find the average for a data set, denoted $\bar{x}$.

1. **Add up all the numbers in the data set.**
2. **Divide by the number of numbers in the data set, *n*.**

For example, player salary data for the 2001–2002 season is shown in Table 5-2 for the 13 players on the Los Angeles Lakers roster (excluding those who were released early in the season).

Table 5-2	Salaries for Los Angeles Lakers NBA Players, 2001–2002 Season
Player	*Salary ($)*
Shaquille O'Neal	$21,428,572
Kobe Bryant	$11,250,000
Robert Horry	$5,300,000
Rick Fox	$3,791,250
Lindsey Hunter	$3,425,760
Derek Fisher	$3,000,000
Samaki Walker	$1,400,000
Mitch Richmond *	$1,000,000
Brian Shaw *	$963,415
Devean George	$834,250
Mark Madsen	$759,960
Jelani McCoy	$565,850
Stanislav Medvedenko	$465,850
Total	**$54,184,907**

*without salary cap adjustments

Adding all the salaries, the total payroll for this team is $54,184,907. Dividing by the total number of players (*n*=13) gives an average salary of $4,168,069.77. That's a pretty nice average salary, isn't it? But notice that Shaquille O'Neal is at the top of this list, and in that year, his salary was the highest in the entire league. If you take the average salary of all of the Lakers players besides Shaq, you would get an average of $32,756,335 ÷ 12 = $2,729,694.58. This is still a hefty amount, but one that's significantly lower than the average salary of all players including Shaquille O'Neal. (Of course, fans would argue that this merely shows how important he is to the team. And this issue is but the tip of the iceberg of the never-ending debates that sports fans love to have about statistics.)

Another word that's used for average is the word *mean*.

So, for the 2001–2002 season, the average salary for the Lakers was about $4.2 million. But does the average always tell the whole story? In some cases, the average may be a bit misleading, and this is one of those cases. That's because every year, a few top-notch players (like Shaq) make much more money than anybody else (and, like Shaq, they also tend to be taller than anyone else, by the way). These are called *outliers* (numbers in the data set that are extremely high or extremely low, compared to the rest of the data). Because of the way the average is calculated, outliers that are high tend to drive the average upward (just like Shaq's salary did in the preceding example). Similarly, outliers that are extremely low tend to drive the average downward.

Remember in school when you took an exam, and you and most of the rest of the class did badly, while a couple of the nerds got 100? Remember how the teacher didn't change the grading scale to reflect the poor performance of most of the class? Your teacher was probably using the average, and the average in that case didn't really represent the true center of the students' scores.

What can you report, other than the average, to show what the salary of a "typical" NBA player would be or what the test score of a "typical" student in your class was? Another statistic that is used to measure the center of a data set is called the median. The median is still an unsung hero of statistics in the sense that it isn't used nearly as often as it should be, although people are beginning to report it more and more nowadays.

Splitting salaries down the median

The *median* of a data set is the value that lies exactly in the middle. Here are the steps for finding the median of a data set:

1. **Order the numbers from smallest to largest.**

2. **If the data set contains an odd number of numbers, choose the one that is exactly in the middle.**

   This is the median.

3. **If the data set contains an even number of numbers, take the two numbers that appear exactly in the middle and average them to find the median.**

The salaries for the Los Angeles Lakers during the 2001–2002 season (refer to Table 5-2) are already ordered from smallest (starting at the bottom) to largest (at the top). Because the list contains the names and salaries of 13 players, the middle salary is the seventh one from the bottom (or top), or the salary of Samaki Walker, who earned $1.4 million that season from the Lakers. This is the median.

This median salary for the Lakers is well below the average of $4.2 million for this team. But because the average Laker salary includes outliers (like the salary of Shaquille O'Neal), the median salary is more representative of the middle salary for the team. (Notice that only 3 players earned more than the average Laker salary of $4.2 million, while 6 players earned more than the median salary of $1.4 million.) The median isn't affected by the salaries of those players who are way out there on the high end, the way the average is. (By the way, the lowest Lakers' salary for the 2001–2002 season was $465,850 — a lot of money by most people's standards but mere peanuts compared to what you think of when you think of an NBA player's salary!)

The U.S. Government often uses the median to represent the center with respect to its data. For example, U.S. Census Bureau reported that in 2001, the median household income was $42,228, down 2.2% from the year 2000, when the median household income was $43,162.

Interpreting the center: Comparing means to medians

Now suppose you're part of an NBA team trying to negotiate salaries. If you represent the owners, you want to show how much everyone is making and how much money you're spending, so you want to take into account those superstar players and report the average. But if you're on the side of the players, you would want to report the median, because that's more representative of what the players in the middle are making. Fifty percent of the players make a salary above the median, and 50% of the players make a salary below the median. That is why they call it the median — like the median of an interstate highway, it's the point in the exact middle.

A *histogram* is a type of graph that organizes and displays numerical data in picture form, showing groups of data and the number or percentage of the data that fall into each group. (See Chapter 4 for more information on histograms and other types of data displays.) If the data have outliers on the upper end, the histogram of the data will be *skewed to the right,* and the mean will be larger than the median. (See the top histogram in Figure 5-1 for an example of data that is skewed to the right.) If the data have outliers on the lower end, the histogram of the data will be *skewed to the left,* and the mean will be smaller than the median. (The middle histogram in Figure 5-1 shows an example of a histogram that shows data that is skewed to the left.) If the data are *symmetric* (have about the same shape on either side of the middle), the mean and the median will be about the same. (The bottom histogram in Figure 5-1 shows an example of symmetric data in a histogram.)

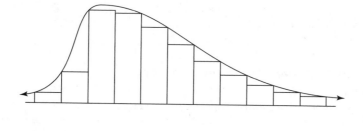

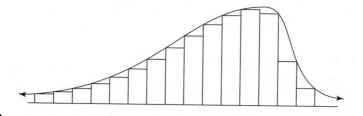

Figure 5-1:
Data
skewed to
the right;
data
skewed to
the left; and
symmetric
data.

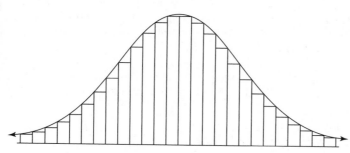

The average (or mean) of a data set is affected by outliers, but the median is not. If someone reports the average value, also ask for the median, so that you can compare the two statistics and get a better feel for what's actually going on in the data and what's truly typical.

Accounting for variation

Variation always exists in a data set, regardless of which characteristic you're measuring, because not every individual is going to have the same exact value for every variable. Variability is what makes the field of statistics what it is. For example, the price of homes varies from house to house, from year to year, and from state to state. Household income varies from household to household, from country to country, and from year to year. The number of passing yards a quarterback achieves in a game varies from player to player, from game to game, and from season to season. The amount of time that it takes you to get to work each day varies from day to day. The trick to dealing with variation is to be able to measure that variability in a way that best captures it.

Knowing what standard deviation means

By far the most commonly used measure of variability is the standard deviation. The *standard deviation* represents the typical distance from any point in the data set to the center. It's roughly the average distance from the center, and in this case, the center is the average. Most often, you don't hear a standard deviation given just by itself; if it's reported (and it's not reported nearly enough) it will probably be in the fine print, usually given in parentheses, like "(*s* = 2.68)."

The standard deviation *of an entire population of data* is denoted with a Greek letter σ. The standard deviation *of a sample from the population* is denoted with the letter *s*. Because most of the time the population standard deviation isn't a value that's known, any formulas involving the standard deviation would leave you high and dry without something to plug in for it. But, never fear. When in Rome, do as the Romans do, right? So when dealing with statistics, do as the statisticians do — whenever they are stuck with an unknown value, they just estimate it and move on! So *s* is used to estimate σ in cases where σ is unknown.

In this book, when I use the term *standard deviation,* I mean *s,* the sample standard deviation. (If and when I refer to the population standard deviation, I let you know!)

Calculating the standard deviation

The formula for standard deviation is $s = \sqrt{\dfrac{\sum (x - \bar{x})^2}{n - 1}}$.

To calculate the sample standard deviation, *s,* do the following steps:

1. **Find the average of the data set.**

   To find the average, add up all the numbers and divide by the number of numbers in the data set, *n.*

2. **Take each number and subtract the average from it.**

3. **Square each of the differences.**

4. **Add up all of the results from Step 3.**

5. **Divide the sum of squares (found in Step 4) by the number of numbers in the data set, minus one (*n* – 1).**

6. **Take the square root of the number you get.**

Statisticians divide by $n - 1$ instead of n in the formula for s so that the sample standard deviation has nice properties that work out with all of their theory. (Believe me, that's more than you want to know about *that* issue.) For example, dividing by $n - 1$ makes sure that the standard deviation isn't *biased* (off target) on average. In case you weren't confused enough already, here's more: If you do ever get the entire population of data and you want to find the population standard deviation, σ, use the same formula as the one for s, except do divide by n, not $n - 1$!

Look at the following small example. Suppose you have four numbers: 1, 3, 5, and 7. The mean is $16 \div 4 = 4$. Subtracting the mean from each number, you get $(1 - 4) = -3$, $(3 - 4) = -1$, $(5 - 4) = +1$, and $(7 - 4) = +3$. Squaring each of these results, you get 9, 1, 1, and 9. Adding these up, the sum is 20. In this example, $n = 4$, and therefore $n - 1 = 3$, so you divide 20 by 3 to get 6.67. Finally, you take the square root of 6.67, which is 2.58, and that is the standard deviation of this data set. So for the data set 1, 3, 5, 7, the typical distance from the mean is 2.58.

Because calculating the standard deviation involves many steps, in most cases, you will probably have a computer calculate it for you. But knowing how to calculate the standard deviation helps you better interpret this statistic and can help you figure out when the statistic may be wrong.

Interpreting the standard deviation

Standard deviation can be difficult to interpret as a single number on its own. Basically, a small standard deviation means that the values in the data set are close to the middle of the data set, on average, while a large standard deviation means that the values in the data set are farther away from the middle, on average.

A small standard deviation can be a goal in certain situations where the results are restricted (for example, in product manufacturing and quality control). A particular type of car part that has to be centimeters in diameter to fit properly had better not have a very big standard deviation. A big standard deviation in this case would mean that lots of parts end up in the trash because they don't fit right; either that, or the cars will have problems down the road.

In situations where you just observe and record data, a large standard deviation isn't necessarily a bad thing; it just reflects a large amount of variability in the group that is being studied. For example, if you look at salaries for everyone in a certain company, including everyone from the student intern to the CEO, the standard deviation could be very large. On the other hand, if you narrow the group down by looking only at the student interns or only at the corporate executives, the standard deviation will be smaller, because the individuals within each of those two groups have salaries that are less variable.

Watch for the units when determining whether a standard deviation is large. For example, a standard deviation of 2 in units of years is equivalent to a standard deviation of 24 in units of months. Also look at the value of the mean when putting standard deviation into perspective. If the average number of Internet newsgroups that a user posts to is 5.2, and the standard deviation is 3.4, that's a lot of variability, relatively speaking. But if you're talking about the age of the newsgroup users, where the mean is 25.6 years, a standard deviation of 3.4 would be comparatively smaller.

Another way to interpret standard deviation is to use it in conjunction with the mean to describe where most of the data are. If the data are distributed in a bell-shaped curve (with lots of data close to the middle, with fewer values as you move away from the middle) you can use something called the empirical rule to interpret the standard deviation. (See Chapter 4.) The *empirical rule* says that about 68% of the data should lie within one standard deviation of either side of the mean; about 95% of the data should lie within two standard deviations of the mean, and about 99% of the data should lie within three standard deviations of the mean.

In a study of how people make friends in cyberspace using newsgroups, for example, the age of the users of an Internet newsgroup was reported to have a mean of 31.65 years, with a standard deviation of 8.61 years. The data were distributed in a bell-shaped curve. According to the empirical rule, about 68% of the newsgroup users had ages within 1 standard deviation (8.61 years) of the mean (31.65 years). So, about 68% of the users were between ages 31.65 − 8.61 years and 31.65 + 8.61 years, or between 23.04 and 40.26 years. About 95% of the users were between the ages of 31.65 − 2(8.61), and 31.65 + 2(8.61), or between 14.43 and 48.87 years. Finally, about 99% of the Internet users' ages were between 31.65 − 3(8.61) and 31.65 + 3(8.61), or between 5.82 and 57.48 years. (For more applications of the empirical rule, see Chapter 8.)

Most people don't bother trying to account for 99% of the values in a data set; they're usually happy with 95%. Going out one more standard deviation on either side of the mean just to pick up an extra 4% of the data (99% − 95%) doesn't seem worthwhile to many people.

Understanding the properties of the standard deviation

Here are some properties that can help you when interpreting a standard deviation:

- ✔ The standard deviation can never be a negative number. (That's because of how it's calculated and the fact that it measures a distance; distances are never negative numbers.)

- ✔ The smallest possible value for the standard deviation is 0, and that happens only in contrived situations where every single number in the data set is exactly the same (no deviation).

✔ The standard deviation is affected by outliers (extremely low or extremely high numbers in the data set). That's because the standard deviation is based on the *distance* from the mean. And remember, the mean is also affected by outliers.

✔ The standard deviation has the same units as the original data.

Lobbying for the standard deviation

The standard deviation is something that is not reported very often in the media, and that's a real problem. If you find out only where the center of the data is without some measure of how variable those data are, you have only part of the story. In fact, you could be missing the most interesting part of the story. Variety is the spice of life, yet without an indication of how diverse or varied the data are, you're not being told how spicy the data are.

Without knowing the standard deviation, you can't get a handle on whether all the data are close to the average (as are the diameters of car parts that come off of a conveyor belt when everything is operating correctly) or whether the data are spread out over a wide range (as are the salaries of NBA players). If someone told you that the average starting salary for someone working at Company Statistix is $70,000, you may think, "Wow! That's great." But if the standard deviation for starting salaries at Company Statistix is $20,000, by using the empirical rule assuming that the distribution of salaries is bell-shaped, you could be making anywhere from $30,000 to $110,000 (that is, $70,000, plus or minus two standard deviations, each worth $20,000). Company Statistix has a lot of variation in terms of how much money you can make, so the average starting salary of $70,000 isn't as informative in the end, is it? On the other hand, if the standard deviation was only $5,000, you would have a much better idea of what to expect for a starting salary at that company.

Without the standard deviation, you can't compare two data sets effectively. What if the two sets of data have about the same average and the same median; does that mean that the data are all the same? Not at all. For example, the data sets 199, 200, 201, and 0, 200, 400 both have the same average, which is 200, and the same median, which is also 200. Yet they have very different standard deviations. The first data set has a very small standard deviation compared to the second data set.

Journalists often don't report the standard deviation. The only reason I can think of for this is that people must not ask for it — perhaps the public just isn't ready for the standard deviation yet. But reference to the standard deviation may become more commonplace in the media as more and more people discover what the standard deviation can tell them about a set of results. And in many workplaces, the standard deviation is frequently reported and used, because this statistic is a standard and well-accepted way of measuring variation.

Being out of range

Many times, the media will report the range of a data set as a way to measure the variability. The *range* is the largest value in the data set minus the smallest value in the data set. The range is easy to find; all you do is put the numbers in order (from smallest to largest) and do a quick subtraction. Maybe that's why the range is used so often; it certainly isn't because of its interpretative value.

The range of a data set is almost meaningless. It depends on only two numbers in the data set, both of which could reflect extreme values (outliers). My advice is to ignore the range and try to find the standard deviation, which is a more informative measure of the variability in the data set.

NBA salaries, as expected, have great deal of variability. Salaries for one single team, the Los Angeles Lakers, for the 2001–2002 season are a typical example. Reference Table 5-2 for the salaries of all 13 players on the team. The average salary is $4,168,069.77, and the median is $1,400,000. The salaries range from the highest, which is $21,428,572 (Shaquille O'Neal) to the lowest, $465,850 (Stanislav Medvedenko) with a range of $21,428,572 − $465,850 = $20,962,722. Wow — that's a huge range! It signifies a big difference between the highest and lowest paid players, that's for sure. But does it mean much in terms of the overall variability in salaries for the whole team? Not really. The standard deviation is $5.98 million, which is still a very big number, but because the standard deviation is based on all the salaries of the team (not just the largest and the smallest ones) the standard deviation has much more statistical meaning than the range does.

When you come across summary statistics, look for the standard deviation to get a handle on how much variation is in the data. If it's not available, ask for it, or go to the source (the press release, the journal article, the researchers themselves), where you are sure to find it. Don't put much credibility in the range; it's too rough of an estimate of variability to account for much of anything.

Determining where you stand: Percentiles

Everyone wants to know how they compare to everyone else. In school, what you got on a test mattered less than how your test score compared to the scores of the other kids in the class. Exams such as the GRE and ACT often keep the total number of points the same each year, while student performances vary as each test changes from year to year. So, along with your score, you always get an accounting of what your score means relative to the others who took the same exam with you. In other words, you find out your relative standing in the group.

Understanding percentiles

The most common way to report relative standing is by using *percentiles*. A percentile is the percentage of individuals in the data set who are below you. If you're at the 90th percentile, for example, that means 90% of the people taking the exam with you scored lower than you did. And that also means that 10 percent scored higher than you did, because the total has to add up to 100%. (Everybody taking the test has to show up somewhere relative to your score, right?)

A percentile is *not* a score in and of itself. Suppose your score on the GRE was reported to be the 80th percentile. This doesn't mean you scored 80% of the questions correctly. It means that 80% of the students' scores were lower than yours, and 20% of the students' scores were higher than yours.

Calculating percentiles

To calculate the kth percentile (where k is any number between one and one hundred), do the following steps:

1. **Put all of the numbers in the data set in order from smallest to largest.**

2. **Multiply k percent times the total number of numbers, n.**

3. **Take that result and round it up to the nearest whole number.**

4. **Count the numbers from left to right (from the smallest to the largest number) until you reach the value from Step 3.**

For example, suppose you have 25 test scores, and when they're put in order from lowest to highest, they look like this: 43, 54, 56, 61, 62, 66, 68, 69, 69, 70, 71, 72, 77, 78, 79, 85, 87, 88, 89, 93, 95, 96, 98, 99, 99. Suppose further that you want to find the 90th percentile for these scores. Because the data are already ordered, the next step is to multiply 90% times the total number of scores, which gives $90\% \times 25 = 0.90 \times 25 = 22.5$. Rounding up to the nearest whole number, you get 23. This means that counting from left to right (from the smallest to the largest number in the data set), you go until you find the 23rd number in the data set. That number is 98, and it's the 90th percentile for this data set.

The 50th percentile is the point in the data where 50% of the data fall below that point and 50% fall above that point. You may recognize this under a different name — the median. Indeed, the median is a special percentile; it's the 50th percentile.

A high percentile doesn't always constitute a good thing. For example, if your city is at the 90th percentile in terms of crime rate compared to cities of the same size, that means that 90% of cities similar to yours have a crime rate that is lower than yours, which is not good for you.

Interpreting percentiles

The U.S. government often reports percentiles among their data summaries. For example, U.S. Census Bureau reported the median household income for 2001 was $42,228. The Bureau also reported various percentiles for household income, including the 10th, 20th, 50th, 80th, 90th, and 95th. Table 5-3 shows the values of each of these percentiles.

Table 5-3	U.S. Household Income for 2001
Percentile	**2001 Household Income**
10th	$ 10,913
20th	$ 17,970
50th	$ 42,228
80th	$ 83,500
90th	$ 116,105
95th	$ 150,499

Looking at these percentiles, you can see that the bottom half of the incomes are closer together than are the top half of the incomes. The difference between the 50th percentile and the 20th percentile is about $25,000, whereas the spread between the 50th percentile and the 80th percentile is more like $41,000. And the difference between the 10th and 50th percentiles is only about $31,000, whereas the difference between the 90th and the 50th percentiles is a whopping $74,000.

By looking at these percentiles and how they're distributed among the data, you can tell that this data set, if shown with a histogram, would be skewed to the right. (A *histogram* is basically a bar chart that breaks the data into groups and shows the number in each group. See Chapter 4 for more on histograms.) That's because the higher incomes are more spread out and trail off more than low incomes, which are more clumped together. In this report, the mean wasn't shown because it would have been greatly influenced by those outliers (the households with very high incomes), which would have driven the mean upward, artificially inflating the overall description of household incomes in the United States.

Percentiles do occur in the media and in many public documents; they can yield some interesting information about the data, including how evenly or unevenly the data are distributed, how symmetric the data are, and some important milestones in the data, such as what the median is. Percentiles can also tell you where you (your test score, your income, and so on) stand in a data set. Sometimes, the value of the average isn't important, as long as you know how far above or below average you are. For more information on these other applications of percentiles, see Chapter 8.

No matter what type of data is being summarized or what type of statistics is being used, remember that summary statistics can't tell you everything about the data. But if these statistics are well chosen and they're not misleading, they can tell a great deal of information quickly. Errors of omission can happen, however, so be sure to be on the lookout for some of those lesser-known statistics that can fill in some important clues to the real story behind the data.

Part III
Determining the Odds

The 5th Wave   By Rich Tennant

RICHTENNANT

"Okay — let's play the statistical probabilities of this situation. There are 4 of us and 1 of him. Phillip will probably start screaming, Nora will probably faint, you'll probably yell at me for leaving the truck open, and there's a good probability I'll run like a weenie if he comes toward us."

In this part . . .

Get your dice ready to roll! In this part, you uncover some of the secrets of the gambling scene (and rule number one is to quit while you're ahead!). You also look at the basics of probability so that you know what you're up against when gambling or dealing with any type of chance or uncertainty. And you may be surprised to discover that probability and your intuition don't always mix!

Chapter 6

What Are the Chances? Understanding Probability

· ·

In This Chapter

▶ Using probability in everyday life and in the workplace

▶ Understanding how probabilities work

▶ Seeing how probability can go against your intuition

▶ Making the connection between probability and statistics

· ·

In this chapter, you discover how probability is used in everyday life and in the workplace and explore some of the rules of probability. You also see how probability and intuition don't always mix, find out ways to avoid some common probability misconceptions, and discover what probability has to do with statistics.

Taking a Chance with Probability

Have you ever said, "What are the chances of that happening?" You read, for example, about two tornados hitting the same tiny Kansas town within a 50-year span. You see a friend on a plane to whom you haven't talked in years. You have two flat tires in one day. Your underdog team wins the NCAA basketball championship during March Madness. Strange things happen, and sometimes these events leave you wondering, "What are the odds? Who would have ever predicted this? What's the chance of that ever happening again?" All of these questions have to do with probability.

But probability isn't just about examining the oddities of life (although that admittedly is a fun pastime for those who engage in it). Probability is really about dealing with the unknown in a systematic way, by scoping out the possibilities, figuring out the most likely scenarios, or having a backup plan in case those most likely scenarios don't happen.

Life is a sequence of unpredictable events, but probability can be used to help predict the likelihood of certain events occurring. Here are some of the more mundane ways that probability may cross your path on a daily basis:

- ✔ The weather reporter predicts an 80% chance of rain today, so you decide to wear your raincoat to work.

- ✔ You know based on experience that going slightly over the speed limit increases your chances of hitting more green lights in a row on your way to work (as long as you don't get a ticket doing it).

- ✔ On your way to work, you wonder whether your assistant, Bob, is going to call in sick today, because it's Friday, and he takes about 75% of his sick days on Fridays. (You also ponder the chance that Bob will tell you he's found another job, an event that has a much lower chance of occurring, you suppose.)

- ✔ You buy a lottery ticket on your lunch hour because "Someone's got to win, and it may as well be me!" (By the way, your chances of winning the jackpot are 1 in 89 million this time, so don't hold your breath.)

- ✔ On TV, you hear about the latest health report that says that if you take a small power nap during the day, you'll reduce your chances of insomnia by 35%. (You fall asleep during the rest of the report.)

- ✔ You end the day by watching your favorite baseball team win another game, and you dream about the chances of winning the World Series.

Probability is also used in virtually every workplace, from marketing companies to investing firms, from government agencies to manufacturing facilities, and from hospitals to restaurants. The following list includes just some of the many examples of how probability is used in the workplace:

- ✔ A small company conducts a survey to find out whether customers like a product enough for the company to offer it on the Home Shopping Network. If the company is right, it can make piles of money; if it's wrong, the company can go broke.

- ✔ A company that makes potato chips has to ensure that the bags are being filled to proper specifications: too few chips, and they'll get in trouble for misrepresenting their product; too many chips, and they'll lose profits. They sample bags of chips and based on those samples, figure out the probability that something is wrong with the machines.

- ✔ Mr. I.M. Hopeful decided to explore the idea of running for governor, but before he goes to the trouble of raising the millions needed to run a campaign, he conducts a poll to determine his chances of winning an election.

✔ A pharmaceutical company has a new drug for high blood pressure. Based on the clinical trials on volunteers, the company determines the probability that someone taking the drug will improve his or her condition and/or develop certain side effects.

✔ A genetics engineer uses probabilities to predict genetic patterns and outcomes in a variety of areas, from designing new crops to identifying hereditary diseases early in a person's life.

✔ A restaurant manager thinks about probability in terms of when and how many customers will come into his restaurant at a given time. He then tries to prepare accordingly.

✔ A stock broker uses probability in her decision-making every day. She constantly wonders whether a given stock goes up or down, whether she should buy or sell, and what she should tell her clients.

Gaining the Edge: Probability Basics

Probability is everywhere, yet it can be hard to understand at times, because it can seem counterintuitive. The first step in gaining the edge on probability is to understand some basic rules of probability and how these rules are applied. When statisticians talk about probability, they talk about the probability of an *outcome*, which is one particular result of a random process being studied. What's a *random process*, you ask? It's any process for which the outcome is not set in stone, but can vary in a random way. For example, if you roll a six-sided die one time, the outcome (the number on the side facing up) will be one of six possible numbers: 1, 2, 3, 4, 5, or 6.

Getting the rules down

Consider the following basic rules of probability:

✔ The probability of an outcome is the percentage of times that the outcome is expected to happen. This can often be calculated by taking the number of ways that the outcome can happen divided by the total number of possible outcomes. For example, the probability of the number 1 appearing when a single die is rolled is 1 out of 6, or ⅙ (or 16.7%).

✔ Every probability is a number (a percentage) between 0% and 100%. (Note that statisticians often express percentages as proportions — numbers between 0 and 1.) If an outcome has a probability of 0%, it can *never* happen, no matter what. If an outcome has probability of 100%, it *always* happens, no matter what. Most probabilities are neither 0% nor 100%, but fall somewhere in between.

✔ The sum of the probabilities of all possible outcomes is 1 (or 100%).

✔ To get the probability of obtaining one of a set of outcomes, you add up the probabilities of each outcome individually. For example, the probability of rolling an odd number (1, 3, or 5) on a single die is the sum of the probabilities of rolling a 1, a 3, and a 5: ⅙ + ⅙ + ⅙ = ½, or 50%.

✔ The *complement* of an event is all possible outcomes *except* those that make up the event. The probability of the complement of an event is 1 minus the probability of the event. For example, rolling a 1, 2, 3, 4, or 5 is the complement of rolling a 6 on a single die, so the probability of rolling either a 1, 2, 3, 4, or 5 is 1 minus the probability of rolling a 6, or 1 − ⅙ = ⅚.

When the complement of an event is complicated, it's often easier to find the probability of the event itself, and take 1 minus the calculated probability. Why take 1 minus this probability? Because the sum of the probabilities of all the outcomes is 1, so the probability of the complement of an event plus the probability of the event must be 1.

Rolling the dice

In the gambling game of craps, two dice are rolled, and the number 7 plays an important role in this game. In craps, each outcome is composed of the two numbers on the dice (for example, the combination 6, 2 is one outcome). The numbers on the two dice are added together to get the sum. (See Table 6-1.) The sum of 7 is the sum that happens most often, and, therefore, has the highest probability of occurring. The shooter (the person rolling the dice) rolls the dice and whatever he/she gets is called the *come-out* roll (for example a 6, 2 combination makes a sum of 8 for the come-out roll). If the come-out roll sums up to 7, the shooter is done with his turn, and everyone that placed a bet loses. If the come-out roll does not sum up to 7, the shooter keeps rolling the dice until either a sum of 7 appears or the sum that showed on the come-out roll appears (in this case, 8). Anyone around the table can bet that a sum of 7 will or won't come up before the sum of the come-out roll comes up again. And that's why everyone at the craps table gets so excited and cheers on the shooter. They're hoping that the shooter will bring them good luck and roll the combinations they're betting on.

You can use the rules of probability listed in the preceding section to look at the outcomes of the sum of two dice and assign probabilities to them. Do you know which sum(s) have the second highest probability of occurring?

When two dice are rolled, each die has six possible results; together, these six possible results on each die yield 36 (6 × 6) possible combinations of two numbers, or 36 possible pairs. Because in this example, an outcome is the

sum of the numbers obtained from the two rolled dice, you have eleven different possible outcomes, which range from 2 (that is, 1 + 1) to 12 (that is, 6 + 6),
and everything in between. Table 6-1 shows the 36 possible results of the dice
rolls, as well as the 11 different sums.

Table 6-1										Outcomes for the Sum of Two Dice	
Result of Dice Roll	**Sum**	**Result of Dice Roll**	**Sum**	**Result of Dice Roll**	**Sum**	**Result of Dice Roll**	**Sum**	**Result of Dice Roll**	**Sum**	**Result of Dice Roll**	**Sum**
1, 1	2	2, 1	3	3, 1	4	4, 1	5	5, 1	6	6, 1	7
1, 2	3	2, 2	4	3, 2	5	4, 2	6	5, 2	7	6, 2	8
1, 3	4	2, 3	5	3, 3	6	4, 3	7	5, 3	8	6, 3	9
1, 4	5	2, 4	6	3, 4	7	4, 4	8	5, 4	9	6, 4	10
1, 5	6	2, 5	7	3, 5	8	4, 5	9	5, 5	10	6, 5	11
1, 6	7	2, 6	8	3, 6	9	4, 6	10	5, 6	11	6, 6	12

You can use the first rule of probability (see the "Getting the rules down" section) to calculate the probabilities for each of the 11 possible sums. A list of all
outcomes and their probabilities is called a _probability model_. For example, a
sum of 7 can happen in 6 different ways: (1, 6), (2, 5), (3, 4), (4, 3), (5, 2), and
(6, 1). With 36 possible combinations for the two dice, the probability that the
sum is 7 is 6 ÷ 36, or $\frac{1}{6}$. Similarly, you can figure out the probabilities of getting
sums 2 through 12. The probability model for the sum of two dice is shown in
Table 6-2. You can see that two sums have the second-highest probability of
$\frac{5}{36}$; these are the sums on either side of 7 (6 and 8). Note that the sum of all of
the probabilities in Table 6-2 is equal to 1. Also note that the probabilities
steadily increase as the sum of the dice goes from 2 to 3 to 4, 5, 6, and peaks
out when the sum of the dice is 7 (that's because the number of combinations
that can result in a sum of 7 is higher than for any other sum). The probabilities steadily decrease again as the sum goes from 8 to 9, and so on, up to 12.

Table 6-2	Probability Model for the Sum of Two Dice
Sum of Dice	**Probability**
2	1/36
3	2/36

(continued)

Table 6-2 *(continued)*

Sum of Dice	Probability
4	3/36
5	4/36
6	5/36
7	6/36
8	5/36
9	4/36
10	3/36
11	2/36
12	1/36

Payouts in any gambling game are based on probabilities. In craps, for example, you can make side bets on what the sum is going to be for any given roll. If you bet that on a given roll the shooter will come up with a sum of 2 and that actually happens, you'll win more than if you bet that on a given roll the sum of 8 will come up. Why? Because getting a sum of 2 on two dice is much less likely to happen than getting a sum of 8 on two dice, according to Table 6-2. That's why they call it gambling. (For more on probability and gambling, see Chapter 7.)

The probabilities for the sum of two dice were fairly straightforward to calculate. However, other probabilities can be more involved, for example the probabilities for different poker hands such as a full house, straight flush, or two pairs. What's important to remember, though, is that the ranking of the hands in poker is directly related to the probability of getting that hand; the highest hand in poker is a royal flush (10, Jack, Queen, King, and Ace, all of the same suit). The reason the royal flush is the highest hand is because it's the one with the lowest probability of occurring.

Models and simulations

Not all probabilities can be calculated using math. In cases when math won't work to calculate a probability, other methods can be used to estimate probabilities or to use known probabilities to make predictions about the world. For example, complicated computer models are used to predict the probability of a hurricane hitting the U.S. coast, and if so, when and where. These computer models are based on data from the behavior of past hurricanes, as well as current weather conditions and other variables. Scientists put the information into a sophisticated mathematical model that tries to predict what the hurricane will do. Work remains to be done in this area, but progress is being made all the time. Models like this would save lives, property, and millions of dollars in damage if people were able to know ahead of time what to expect and prepare accordingly.

Other models are based on observational data. The U.S. Census Bureau's American Community Survey surveyed households in Columbus, Ohio, in 2001 to get an idea of the makeup of the community. One of the characteristics examined was household composition (married couples, other families, people who live alone, and other non-family households). The data are summarized in Figure 6-1. These statistics from a *sample* of households can serve as a probability model for the makeup of *all* households in Columbus, Ohio, in 2001.

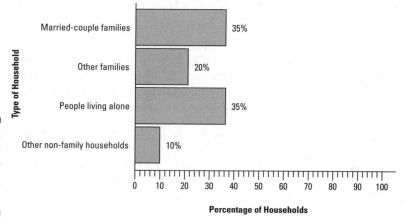

Figure 6-1:
Household make up for Columbus, Ohio, 2001.

For example, because 35% of the sampled households were married-couple households, you can say that the probability that a randomly selected household in Columbus is a married-couple household is 35%. You can also use the rules of probability to make other statements about the households in Columbus in 2001. For example, what's the probability that a randomly selected house contains any type of family? That would be the sum of the probabilities of selecting a married couple household (35%) and a household that falls into the "other family" category (20%). So, the probability that a randomly selected household in Columbus, Ohio, in 2001 contains a family is 35% + 20% = 55%. (Therefore, the probability of selecting a non-family household is 100% – 55% or 45%.)

The probability model in Figure 6-1 shouldn't be used for other communities outside of Columbus, Ohio, because in this survey, the sample of households was selected only from Columbus. Using these data to discuss a population other than the one from which the sample was drawn would be invalid. (See Chapter 16 for more on surveys and what they can and can't say about populations.)

Simulations are another way to estimate probability when a formula isn't possible. In a *simulation,* a process is repeated over and over again under the exact same conditions (usually using a computer), and the outcomes are recorded each time. The probability of any outcome is estimated by the percentage of times the outcome occurred in the simulations. For example, a sports fan with too much time on his hands simulated thousands of NCAA tournaments on his computer and used these simulations to predict that Duke would win the NCAA basketball championship in 2002 with a probability of over 95%. As luck would have it, the prediction turned out to be wrong (Duke was eliminated early in the tournament) proving that the only thing you can be certain about is uncertainty.

Interpreting Probability

A probability can be interpreted in two ways: as a short-term chance, or as a long-term percentage. In the short term, the probability of an event is the percentage chance that the event is going to happen on the next try. For example, your meteorologist may tell you that the probability of rain tomorrow is 40%. Or a baseball player's batting overall average is 0.291 (meaning he has on average a 29.1% chance of getting a hit the next time he comes up to bat).

Probability also means the percentage of times that an event will happen in the long run (over a long period of time with repeated trials under the same conditions). So a 40% chance of rain tomorrow can be taken to mean that if you look at data from a large number of days similar to the type of day tomorrow is supposed to be, it rained on 40% of those days. The baseball player's 0.291 batting average can be interpreted as the proportion of times that he gets a hit averaged over many times at bat (in this case he's expected to hit the ball 291 times out of 1,000 times at bat).

Avoiding Probability Misconceptions

The basic rules of probability seem pretty straightforward, but probability can often be counterintuitive. This section gives you some of the more common misconceptions people have about probability.

Looking more likely

If you were to write down a sequence of what you would think the outcomes of six flips of a *fair coin* (that is, a coin that hasn't been tampered with) would look like, you probably wouldn't think of writing down something like

HTTTTH (where "H" means "heads" and "T" means "tails"), because that doesn't look very "random." However, this exact sequence of heads and tails has the same chance of happening as any other exact sequence. That's because the probability of getting a head is the same as the probability of getting a tail on each individual toss. Now, if you were to compare the probability of getting two heads (out of six tosses) with the probability of getting six heads (out of six tosses), you'd get different values. The probability of getting two heads (out of six tosses) is higher because you have more ways of accomplishing it, instead of having to get a head every single time.

With the lottery, the sequence 1, 2, 3, 4, 5, 6 has the same chance of winning as any other combination of six numbers, even though it doesn't look like it can *ever* occur. This fact makes you realize that all of the other combinations are just as *unlikely* to be chosen as this combination is. However, if you bet on this combo and win, you probably won't have to split the winnings with anyone.

Predicting long or short term

Probability works well for predicting long-term behavior, but it doesn't work well for predicting outcomes in the short term. In the long-term, you know that unless the event has a probability of 0, it will happen sometime, and depending on how big the probability is, you can even get some idea of how long you can expect to wait. But you won't know exactly *when* the event will happen. That's what makes probability so interesting and what keeps gamblers coming back over and over again.

For example, if I flip a fair coin six times and get six heads in a row, what do you think the outcome of the next flip should be, a head or a tail? You may think I'm due to get a tail, so getting a tail now should have a higher chance of happening. But in fact, the probability of getting a tail on the next flip is still ½, the same as it was for each of the previous flips. You know that if the coin were flipped a large number of times that you can expect about 50% of the outcomes to be heads and 50% to be tails. But you can't predict *when* those heads or tails will appear on any given flip of the coin. (So even though it seems like a tail is due, the probability of getting a head or a tail *on this next flip* is still 50%.) Eventually, tails will start coming up, but you can't say when.

Thinking 50-50

One common misconception is to think that every situation with two possible outcomes is a "50-50" situation (in other words, a 50% chance that each of the two outcomes will occur, just as it was with the outcome of a single toss of a fair coin). Many people think that just because two outcomes are possible, each outcome must have a one out of two chance of occurring, but that's very often not the case. Not every situation is like a fair coin toss. Many situations have a higher probability of one outcome over the other.

For example, think of a computerized "walk/don't walk" sign on the crosswalk of a busy street. Is that sign going to say "walk" exactly 50% of the time? No. When the street is busy, the light will stop traffic less often, and pedestrians will have to wait longer between opportunities to cross the street. Using a sports example, think of a basketball player standing at the free throw line. Are her chances of making the basket 50-50? (After all, she either makes it or she doesn't.) Her chances are 50-50 only if her overall free throw percentage is 50% over many tries. Most likely, it's something higher than that.

Interpreting rare events

Probability can become a topic of controversy, especially in the case of rare events. A rare event has a small probability of happening, but what does that mean? It means that for any single situation or person, the event is unlikely to occur, yet if given enough repetitions of the situation over a long enough period of time or with enough people, the event is bound to happen to somebody, somewhere, sometime. This comes into play in situations where you have a cluster of people with a rare disease in one town, and you need to figure out whether something caused this to happen (the air, the water, the soil, and so on) or whether this just occurred by chance (something most people don't consider).

Because it doesn't seem very likely that a rare event would actually occur, people naturally want to blame the occurrence on something. In some cases, they'd be right; in other cases, this is just a phenomenon of random chance. Do three years in a row of rising average temperatures indicate global warming? If a dairy farm had two cows both give birth to two-headed calves in the same season, does that mean their cows have a terrible problem? How many tire blowouts should it take to constitute a tire recall? Looking at something after the fact and saying, "What was the chance of that happening here?" is different than before the fact knowing that the same event is bound to happen somewhere, sometime.

For example, if you flip a fair coin long enough, eventually, you come across a long string of heads, just by chance. That's supposed to happen sometime. And because the coin was fair, you couldn't blame it on anything but chance. However, the media may try to establish a pattern when they see two or more occurrences of an event, such as child abductions across the country, nightclub fires, or occurrences of a rare disease in the same city. I'm not saying these shouldn't be investigated for possible causal problems, but I am saying that the media needs to be aware that sometimes, events just happen in clumps by chance, with no big story behind it. It's also interesting to note that people view the probability of a rare event differently depending on whether the rare event is a good thing, like winning the lottery ("It's got to happen to somebody, so it may as well be me!") or a bad thing, like getting struck by lightning at a golf tournament ("That's a million-to-one shot. That can never happen to me!"). This may just be human nature. ***Mental note:*** Human nature doesn't correspond to the laws of probability.

To avoid some of the more common probability misconceptions, keep the following in mind:

- Probability isn't effective in predicting short-term behavior. It is effective when predicting long-term behavior.

- In the case where only two outcomes are possible, each outcome doesn't necessarily have a 50% chance of occurring.

- If a cluster of rare events occurs somewhere, it may have happened due to chance and no other reason. Rare events are going to happen to somebody, somewhere, sometime, given enough people and time.

- You can't be "on a roll" if a process is being repeated over and over under the same conditions (as in a gaming situation). Probability has no memory.

- Sequences of outcomes that "look more random" oftentimes have the same probability as sequences that don't "look as random." For example, you may think that HTTTTH has a smaller chance of occurring than HTTTHT does because it doesn't "look as random." In fact, they each have the same probability of occurring, because each of the outcomes contains four tails and two heads (and the order doesn't matter when calculating the probability here).

Connecting Probability with Statistics

You may be thinking, "Probability is interesting, but what does it have to do with statistics?" Good question. It may not seem obvious, but probability and statistics fit together like a hand in a glove. Data are collected from a sample of individuals, and then statistics are calculated to summarize those data. But you don't stop there. The next step is to use these statistics to make some sort of prediction, generalization, conclusion, or decision about the population that this sample came from. That's where probability comes in.

Estimating

Data are often collected in order to help estimate population proportions or averages. For example, doctors estimate the chance of someone having a heart attack by first gathering information about a patient's weight, body mass index, age, gender, genetic background, diet, exercise level, and so on. Then they compare the information to data that have been collected from a sample of people who have similar characteristics to the patient, and they come up with the patient's probability (or risk level) of having a heart attack in a given period of time. Engineers estimate the average number of cars that will be using a certain section of the interstate highway during rush hour by recording traffic data using technology in the pavement. After the data are collected, probability is used to determine how much the sample information is likely to vary from sample to sample, day to day, hour to hour, and so on.

Predicting

Statistics are involved in helping to make predictions of all kinds — everything from weather predictions and population size projections to the spread of disease or the future values of the stock market. Data are collected over a period of time and are analyzed to find a model that not only fits the data well, but also allows for some predictions to be made for the near future. Probability helps people using the models to assess how accurate those predictions are expected to be, given the data at hand. Probability also helps scientists determine what the most likely scenario is, given the data.

For example, the U.S. Census Bureau makes available its population projections for the total U.S. resident population. You can currently look at population projections all the way to the year 2100. In 2000, the projected population for 2003 was 282,798,000, and as of May 2003, the population of the United States (as shown on the U.S. Census Bureau's Web site) was already 291,065,455 and counting. So, the projection was already off by about 8.3 million people with much of the year left, but that's only 2.8% of the total population in this situation. Estimating the total population size of the United States in the future is a difficult job. It's hard enough to count the number of people living here right now! (By the way, according to Census Bureau, the size of the U.S. resident population projected for the year 2100 is 570,954,000.)

Deciding

Many decision-making processes involve statistics and probability. Medical treatments are often decided in terms of the percentage of people that did well using that treatment, compared to others. The probability that the next person will do well on a treatment would be estimated from the percentage of other patients who did well on that treatment. Most liability forms you need to sign before having surgery outline the possible side effects or complications and give some indication of how often those happen. (See Chapter 17 for more on medical studies.)

Checking quality

Other decisions that involve probability occur during manufacturing processes. Many companies that manufacture items do some sort of quality control; that is, they sample products that come off of the line and assess product quality according to some set of specifications. Probability is used to decide whether and when the manufacturer needs to stop the process due to a problem with the quality of the products. Differences between the sampled product and the specifications may just be due to random variability or an unrepresentative sample; alternatively, these differences can mean that something is wrong with the process. Stopping the process unnecessarily costs money and time, but not stopping the process when it needs to be stopped costs the company in terms of customer satisfaction with the product. So probability is used to make some pretty important decisions in the manufacturing world. (See Chapter 19 for more on quality control.)

When generalizing results from a sample to a population, probability is used to assess the accuracy of those generalizations. Probability is used to determine which conclusion is most likely and why. When making a decision about a situation with an unknown outcome, you use probability to assess the evidence that has been collected, to make a choice based on that assessment, and to know the chance that you made the right or wrong decision. (See Chapter 14 for more information.)

Chapter 7

Gambling to Win

· ·

In This Chapter

▶ Realizing why casinos make money

▶ Understanding the probability behind the games

▶ Taking your money and running: tips for gamblers

· ·

Las Vegas is one of the most exciting hotspots in the world. I have a feeling though, that the cheap all-you-can-eat buffets and the majestic Roman gladiators that roam Caesar's Palace aren't the main attractions (although I highly recommend both!). Las Vegas is arguably a gambler's paradise; the place to go when you're feeling lucky and want to win big. The fact that the vast majority of folks who gamble in Las Vegas come out losers isn't important to that new crop of potential winners who board the planes each day heading toward Nevada with a feeling of hope and exuberance. After all, someone has to win, right? It may as well be you. While I can't promise that this chapter is going to make you a big winner in Vegas (or any place else you may choose to try a round with Lady Luck), I can say that it will help you understand what you're up against, give you some tips to gain a possible edge, and at the very least, help you not lose quite so much.

Betting on the House: Why Casinos Stay in Business

Casinos are beautiful places; they have exciting atmospheres with brightly colored machines, blinking lights, exciting sounds, happy dealers and servers, and no clocks or windows anywhere in the place (they don't want you to realize how much time you've been spending there). Everything is well thought out, from the layout of the building (to get to any restroom you must go past tons of slot machines) to the patterns chosen for the carpet (the patterns are intentionally busy and hard on the eyes, because the owners don't want you to look down but instead want you to always be looking up at the action and moving toward your next adventure).

The gaming industry has its act down to a science, and those who run gambling houses are very good at what they do. They offer you great entertainment and a chance to win some big money; all you have to do is play. Many people are lured by the chance to win thousands of dollars or a new car with just one pull of a slot machine or just one perfect hand of blackjack. Of course that chance exists, but if everyone won big money, the casinos couldn't afford to stay in business. So they have to find ways to take your money, and they do that by setting game rules that give them a tiny edge during each round of play; by making sure that when someone does win, he or she wins big (so word gets around); and by encouraging you to stay in the casinos as long as possible. They know that the longer you play, the greater their chances of taking your money.

Hundreds of books have been written about how to beat the casinos at most any game offered. Each author wants you to believe that his own strategy is going to make you win big. The truth is, the best these books can do is to help you not lose so much, because the way the games are set up, the house (the casino) always has an advantage. This advantage is smaller with certain games, such as Blackjack, and larger with other games, such as slot machines, which, according to rumors, generate up to 80% of the total profits for some casinos. The most important thing to remember in any gaming situation (*gaming* is the casino industry's softer word for gambling) is to quit while you're ahead. If everyone did that, the casinos would be out of business. But of course, that won't happen because quitting while you're ahead is hard to do.

On the other side of the coin, another good strategy is to cut your losses and quit before you lose too much, instead of figuring that the odds will eventually turn back in your favor, and you'll win all your money back. That approach often turns out to be a lose-lose situation. Casinos bet that you will stay and play no matter what, increasing their chances of eventually getting more and more of your money. Looking at the lavish casinos now being built in Las Vegas, it seems that their bets are paying off.

What can you do to minimize your chance of losing, or of losing too much? Set up some boundaries for yourself before you even walk into the gaming situation (for example, quit playing when you're ahead or down by a certain amount), and then stick with those boundaries. When gambling (excuse me — gaming), quit while you're ahead or before you get too far behind. Set your limits before you start.

Knowing a Little Probability Helps a Lotto

Two of the most important tools you can use in any gaming situation are information about your chances of winning and a true understanding of what those chances mean. Probability can certainly go against your intuition, and you don't want to let your intuition get in the way of keeping your money. (See Chapter 6 for a more in-depth discussion of the use of probability in statistics.)

Here are some of the most common *misconceptions* about probability:

✔ Any situation that has only two possible outcomes is a 50-50 situation (50% chance of winning, 50% chance of losing).

✔ A combination of lottery numbers like 1-2-3-4-5-6 can never win; those numbers aren't random enough.

✔ Buying 100 lottery tickets instead of just one is a great idea; it gives you such a better chance of winning.

✔ If Joe and Sue have three girls already, the chance that they will have a boy next time has to be pretty good.

✔ The longer you play this slot machine, the better chance you have of coming out ahead.

In this section, I break down each of these misconceptions so that you can be informed about the reality of any gaming situation, have a better understanding of what to expect, and plan accordingly. This may ruin some of the magic and excitement that comes with gaming, but then again, it probably explains why you can't find statisticians (certainly none that I know) who are either professional or compulsive gamblers. In fact, Las Vegas won't really mind if statisticians don't hold a conference there any more, because they didn't spend much money in the casinos the last time they were there! (I, myself, play only nickel slot machines, one nickel at a time. At least my $20 will last longer.)

Having a 50-50 chance

You're going to flip a fair coin (that is, a coin that hasn't been tampered with); it has heads on one side, tails on the other. What's the chance that a head will come up? Fifty percent. What's the chance that a tail will come up? Fifty percent. If you were to bet on the outcome of this coin flip, you would have a 50-50 chance of winning (versus losing). Why is that? Because you have two possible outcomes, heads and tails, and each of these outcomes has an equal chance of occurring. You know a couple that is going to have a baby. They, of course, can have either a boy or a girl. Each of these outcomes is equally likely, so the couple's chances of having a girl (versus a boy) is 50-50. On the other hand, if you buy one of 1,000 raffle tickets for a motorcycle, you have two possible outcomes, win or lose the motorcycle. Does that mean that your chance of winning is 50-50? No. Why not? Because you aren't the only person who bought a ticket!

Four rules of probability may help break these ideas down a bit:

- ✔ The probability that a certain outcome will happen is the percentage of times that the outcome is expected to happen in the long term, if the exact same conditions were repeated over and over.

- ✔ Any probability is a number between 0 and 1. A probability of 0 means that the outcome is not even possible. A probability of 1 means that the outcome is certain.

- ✔ All of the probabilities for all possible outcomes must add up to 1. That means the probability that an outcome does *not* happen equals 1 minus the probability that the outcome *does* happen.

- ✔ The probability of an event (a combination of outcomes) is equal to the sum of the probabilities of the individual outcomes that make up the event.

With the coin flip, two outcomes are possible, head or tail. The number of ways a head can occur is 1, and the number of ways a tail can occur is 1. The total number of possible outcomes is 2: head or tail. So, in that case, the chance of a head coming up is ½ or 50%; the same is true for a tail. This is a 50-50 situation.

However, if you look carefully at the first rule, this explains why not everything with two outcomes has a 50-50 chance of happening. You have to look at the number of ways that each of the outcomes can happen. With your motorcycle raffle ticket, you can either win or lose. The number of ways you can win is 1, because the raffle organizers will draw only 1 winning ticket. The number of ways you can lose is 999 because all of the remaining tickets are losers. The total number of outcomes is 1,000. That means your chance of winning is 1 ÷ 1,000 = 0.001, and your chance of losing is 999 ÷ 1,000 = 0.999. Certainly, you have only two possible outcomes regarding your raffle ticket (win or lose), but each of these outcomes does *not* have an equal chance of happening, so this is definitely not a 50-50 situation.

Very few probabilities in life are actually 50-50. To be in a 50-50 situation, you must have only two possible outcomes *and* the probability for each of those outcomes must be the same, that is, 50%. In most situations with two possible outcomes, the two outcomes are not equally likely.

Picking winning numbers

So, you're ready to play the Powerball lottery. You've heard that the jackpot is now up to $200 million — you may buy more tickets than usual this time to increase your chances of winning. And you're ready to pick those numbers. (You have to pick five different numbers between 1 and 53, and then you pick a separate Powerball number between 1 and 42, and this one can be the same as one of the other five that you already chose.) To win the big jackpot, you need to correctly pick (in any order) all of the first five numbers, plus the correct Powerball number. What combination should you pick? Your brother's football number, your mom's birthday, four digits from your Social Security number, your dog's age in months, and the number you saw in your dream last night? These options sound as good as anything else, because any combination that you choose is just as likely to win as any other combination.

This makes good sense until you start thinking about the combination 1-2-3-4-5 with a Powerball number of 6. It seems like this combination should never happen because these numbers don't seem random enough. Well, this is a case in which your intuition can get the best of you. Indeed, this combination has the same chance of being chosen as any other combination that can occur. Take a small example in which the possible numbers to choose from are 1, 2, 3, 4, and you have to pick 2 numbers. The six possible outcomes are 1-2; 1-3; 1-4; 2-3; 2-4; 3-4. Your chance of winning is 1 out of 6. Notice that the combination 1-2 has the same chance of being chosen as any other combination. The same is true for the combination 1-2-3-4-5 with a Powerball of 6. Combinations like 23-16-05-24-18 with a Powerball of 12 may look easier to get, but remember that you have to get every single number exactly right in order to win the big jackpot.

A combination like 1-2-3-4-5 with a Powerball of 6 looks hard to get, but it actually has the same chance of winning as any other combination. What this combination should make you realize is how *small* the chance of winning the jackpot really is. (The actual chances of winning the jackpot with a Powerball such as the one described above are 1 in 120,526,770.)

Lotteries typically call the probability of winning the *odds* of winning, and in this case, these two terms are taken to mean the same thing. However, the way payouts and odds are presented in sports betting (horse races, football games, boxing, and so on) is different from what's described here, and that more complex form of betting odds is beyond the scope of this book.

Buying lottery tickets — less can be more

A Powerball lottery ticket only costs a dollar, and it gives you a chance to win a multi-million dollar jackpot. You figure that someone has to eventually win that jackpot, so you decide you're going to take your shot at it. After all, if you don't play, you can't win. As long as you truly understand your chances of winning *and* losing, buying a few lottery tickets now and then can be cheap fun.

The problem, however, comes when people buy lots of tickets, thinking that their odds are greatly increased by buying more tickets. While holding a hundred tickets (versus only one ticket) does increase your chance of winning a hundred times, you have to realize that the chances of winning big are very small — almost zero. And a hundred times a number that's close to zero is still very close to zero. If you can't afford to lose that hundred dollars (and the odds are overwhelming that you will lose it), don't bet it all on the lottery.

To put the probability of winning the Powerball jackpot into perspective, look at Figure 7-1. This shows the prizes and the chances of winning each prize in a given Powerball lottery. (Most Powerball lotteries offer the same payouts with the same odds.) The gray circles indicate the five balls chosen from 1 to 53 and the black circle indicates the Powerball number. The odds of winning the $3 prize are 1 in 70, or 0.0142 (about 1.5%). Note that this is more like a probability than real odds, but I'll let it slide. As you go up the scale, the winnings increase, but your odds of winning decrease, and they do so exponentially. Matching 4 of the 5 numbers, for example, has odds of 1 in about 12,000; matching 5 of the 5 numbers has odds of 1 in about 3 million; and finally, matching all 5 numbers plus the Powerball number, decreases your odds to one in over 120 million.

You Can Win $10 Million or More!		
Match	**Prize**	**Odds**
⚪⚪⚪⚪⚪ + ⚫	Jackpot	1:120,526,770
⚪⚪⚪⚪⚪	$100,000	1:2,939.677
⚪⚪⚪⚪ + ⚫	$5,000	1:502,195
⚪⚪⚪⚪	$100	1:12,249
⚪⚪⚪ + ⚫	$100	1:10,685
⚪⚪⚪	$7	1:261
⚪⚪ + ⚫	$7	1:697
⚪ + ⚫	$4	1:124
⚫	$3	1:70

Figure 7-1: Payouts and chances of winning for a Powerball lottery.

Why do the odds decrease so quickly just by having to match one more number each time? A small example may help illustrate this. Suppose you have to pick 2 numbers between 0 and 9. If you have to match only one number, your chances are 1 in 10. If you have to match both numbers, your chances drop to 1 in 45. This is because you have 10 possibilities for the first number, times 9 possibilities for the second number (because no repeats are allowed), for a total of 90 possibilities. But now you have to divide those 90 possibilities by 2, because you can have the numbers in either order and still win. (You don't want to count 1-0 and 0-1 as separate combinations, for example.) The odds change dramatically because of this multiplier effect. Having to match the Powerball number in addition to having to match the first five numbers multiplies the odds by an additional 42 times (because you have 42 possible numbers for the Powerball).

The overall chance of winning *any* prize (not just the big prize) is reported by the lottery to be 1 in 36. Because the overall chance of winning is the same as the chance of winning any prize, add up all the probabilities of the prizes, and you get 1 in 36. You're using the fourth rule of probability here (see the "Having a 50-50 chance" section earlier in this chapter).

Before you play any game of chance, always look at the odds or probability of winning. And don't spend more money than you can easily afford to lose. With games that have big payoffs, the chances of winning are always extremely small, and buying many more tickets, or playing many more times, won't increase those chances by enough to justify the added cost. Like the saying goes, the best way to double your money in gambling is to fold it in half and put it in your pocket!

Predicting a boy or a girl

Joe and Sue have already had three girls, and they're expecting their fourth baby soon. They really want to have a boy this time. Friends and family think that their chances of having a boy are higher this time, because they've had three girls in a row. Are they right? This is similar to the people around a craps table who are cheering on the shooter (the person with the dice) because he's on a "hot streak" and can do no wrong. Do winning or losing streaks really exist, and after a certain string of events has happened, does that increase or decrease the chances that the same thing will happen again?

In many situations, especially in gambling, winning or losing streaks don't exist, because each time you play a game of chance, everything resets itself, and the outcome from last time has no effect on the outcome this time or on the outcome next time. In other words, when events are independent of each other, the probability of an event remains exactly the same each time the game is played.

For Joe and Sue, this may be bad news, but the probability of having a boy is still 50-50, the same as it's always been, regardless of whether they have already had three girls. Similarly, if you flip a fair coin three times and get heads each time, you shouldn't expect a tail to have a higher chance of appearing on the fourth toss. The chance is still 50-50.

Probabilities work only in predicting long-term behavior, not short-term behavior. If you know that the chance of heads is 50%, that means if you flip the coin many times, you should expect about half of the outcomes to be heads and half to be tails. However, you can't predict *when* those heads and tails will appear at any given moment. They don't even themselves out as they go along; they only approach their actual probabilities in the long term. This phenomenon is called the *law of averages,* which is further defined in the following section. Many people use this term to explain why winning or losing streaks end, even when these so-called streaks don't even exist.

Trying to win at slots

The old one-armed bandit (that is, the slot machine) is a powerful force. These machines have pans made of a special material that loudly projects the sound of coins hitting the pans. The machines blink and make beeping noises every time you win anything; they even beep when you put your money into them, just to add more of that winning sound to the casino atmosphere. Some say that the *loosest slots* (the ones with the best payouts) are by the entryway of the casino or at the end of the rows. Some say the *tightest ones* are by the blackjack tables because people don't want to be distracted too much by the beeping noise. Casino owners never tell their secrets, so no one can tell for sure, but one thing is certain: Slot machines can take your money very quickly. No skill is needed, and each pull of the handle takes only a few seconds to spend anywhere from five cents (in my case) to thousands of dollars (in the high-dollar slot areas of some casinos). All of that spending and pulling is done in pursuit of the big jackpot that may come on your very next spin.

One of the most common misconceptions about slot machines is that the longer you play them, the greater your chances of winning. This is something that the casinos are betting on. In fact, just the opposite is true, because of the law of averages. The *law of averages* says that in the long run, averages will be close to their expected value. In statistical terms, the *expected value* is a weighted average of the outcomes based on their probabilities.

With any casino game, the house has a slightly higher chance of winning any single round. This means in the long run, you should expect to lose a small amount with a very high chance every single time, and you should expect to win a large amount with a tiny chance every single time. The casino sets their odds and payouts so that, in the long term, everything averages out in their favor, even taking the big jackpots into account.

What this means for you is that you'll end up with a small negative expected value every time you play. And the more you play, the more money you have to expect to lose in the long term, because your overall expected value is the sum of the expected values for each time you play; and each one of those expected values is a negative number. Now you know why the casino gives you free beverages while you're gambling and why you won't see any clocks on the walls, or windows to look out of to watch the changing seasons while you sit on that slot-machine stool. The casinos are betting that you'll forget all about the law of averages as it sneaks up on you while you play.

The only way to beat the slots is to quit while you're ahead, before the law of averages takes over. Remember, the casinos have probability on their side because they're in it for the long term. You can't fool with Mother Nature, and you can't fool with probability. And if you do win, take your money and run!

Part IV
Wading through the Results

The 5th Wave
By Rich Tennant

"WHAT EXACTLY ARE WE SAYING HERE?"

In this part . . .

This part helps you understand the underpinnings of statistics — the information that helps you understand the deeper issues that are working behind the scenes whenever statistics are formulated. You find out how to measure variability from sample to sample, how to come up with a formula to measure accuracy of a statistic, and how to measure where an individual stands with respect to the rest of the population (called a measure of relative standing). All of these topics help build your confidence in interpreting and understanding statistics from the ground up.

Chapter 8

Measures of Relative Standing

The only way to really be able to interpret statistical results is to have something to compare them to, so that you can put the results into some type of perspective. For example, suppose a hypothetical physical therapy student named Rhodie takes a standardized test for physical therapy certification and gets a score of 235. What does 235 mean in this case? Nothing, if that's all you have to go on. You need to be able to put this score into some perspective by determining where it stands relative to the other scores on the test. Is a light bulb that lasts more than 1,200 hours a freak of nature or just a standard light bulb? You can't tell without knowing how long most light bulbs last. Suppose Bob's average exam score at the end of a math course was 78; is that a B or a C? It depends on how his average score compares to the other average scores in his class (and how nice his professor is!).

In this chapter, you discover how to find and interpret the relative standing of individual results; the goal here is to describe where an individual stands, relative to all of the other individuals in the population. In Chapters 9 and 10, I discuss how to find and interpret the relative standing of the results from a sample (for example, the sample mean or the sample proportion). The goal in that case is to determine where your sample mean or sample proportion stands, compared to the population of all possible values of the sample mean or sample proportion.

Straightening Out the Bell Curve

The first step in determining where a particular result stands is to get a listing or picture of all of the possible values that the variable can take on in the population and how often those values occur; this is called a *distribution*.

Many different types of distributions are possible. For example grades for one class (call them Mr. Average's class) are distributed uniformly, with an equal number of scores in each grouping (see the top portion of Figure 8-1) while grades from another class (call them Mr. Mean's class) are distributed in a polarized way, with everyone getting either a very high or a very low score (as in the bottom portion of Figure 8-1). (Most distributions tend to fall somewhere in between.) Notice that, for any distribution, the total of all the percentages has to be 100%, because every value has to appear somewhere on the distribution.

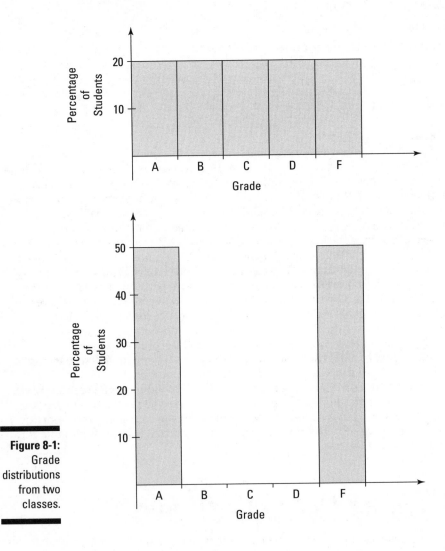

Figure 8-1:
Grade
distributions
from two
classes.

A *bell-shaped curve* describes data from a variable that has an infinite (or very large) number of possible values, and these values are distributed among the population in such a way that when they're plotted in a histogram, the resulting figure has the shape of a bell. This basically means you have a big group of individuals near the middle of the distribution, with fewer and fewer individuals trailing off as you move farther and farther away from the middle in either direction. Many variables in the real world (for example, standardized test scores, lifetimes of products, heights, weights, and so on) have distributions that look like a bell-shaped curve. That makes the bell-shaped curve (sometimes called simply a *bell curve*) important enough to be singled out among all the other possible distributions.

Statisticians call a distribution that has the shape of a bell curve a *normal distribution*. You can see a picture of a normal distribution in Figure 8-2. In this example, the variable is the number of hours that a certain company (call it Lights Out) expects its light bulbs to last. (How would you like to be the person collecting the data to test that little fun fact?)

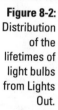

Figure 8-2:
Distribution of the lifetimes of light bulbs from Lights Out.

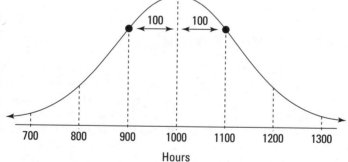

Hours

Characterizing the normal distribution

Every bell-shaped curve (normal distribution) has certain properties. You can use these properties to help determine the relative standing of any particular result in the distribution. The following is a list of properties shared by every normal distribution. These properties are explained in more detail in the following sections.

- ✔ The shape of the curve is symmetric.
- ✔ It has a bump in the middle, with tails going off to the left and right.
- ✔ The mean is directly in the middle of the distribution. The mean of the population is designated by the Greek letter μ.

✔ The mean and the median are the same value, due to symmetry.

✔ The standard deviation represents a typical (almost average) distance between the mean and all of the data. The standard deviation of the population is designated by the Greek letter σ.

✔ About 95% of the values are within two standard deviations of the mean.

Describing the shape and center

A normal distribution is *symmetric*, meaning that if you fold it in half right down the middle, the two halves are mirror images of each other. Because its curve is symmetric, the *mean* (the balancing point) and the *median* (the point where half of the data lie on either side) are equal, and they both occur at the middle of the distribution. The lifetimes of the light bulbs shown in Figure 8-2 have a normal distribution with a mean (and median) of 1,000 hours. (See Chapter 5 for information on the mean and median; see Chapter 4 for more on symmetry.)

Measuring the variability

The shape and the mean aren't the only important characteristics to consider when looking at a distribution. The variability in the values is also extremely important, even though much of the media ignores this characteristic and typically reports only the mean. Referring to Figure 8-2, you can see that the bulk of the light bulbs from Lights Out has a range of lifetimes that vary from under 700 hours to over 1,300 hours, with a good many of the bulbs lasting between 900 and 1,100 hours. As a consumer, do you want that much variability in lifetimes when you buy a package of light bulbs? Maybe not. A competing company (call them Lights Up) is going to try to produce light bulbs with less variability; the lifetime of their light bulbs will still have a mean of 1,000 hours, but this company is able to produce bulbs with more consistent lifetimes, ranging from around 940 to 1,060 hours, with a good many of their light bulbs lasting between 980 and 1,020 hours (see Figure 8-3).

Variability in a distribution is measured and marked off in terms of number of *standard deviations*. (See Chapter 3 for the formula for standard deviation.) On a normal distribution, the standard deviation has a special significance because it's the distance from the mean to a place on the distribution called the *saddle point*. Each normal distribution has two saddle points; each is the same distance from the mean. To find a saddle point, start at the mean and move either right or left until the curvature changes from being an upside-down bowl (concave down) to a right-side-up bowl (concave up).

In Figures 8-2 and 8-3, the saddle points are marked with dots. The standard deviation of the light bulb lifetimes from Lights Out (refer to Figure 8-2) is 100 hours. The standard deviation of the more consistent light bulbs from Lights Up (see Figure 8-3) is 20 hours. (For more information on standard deviation, see Chapter 5.)

Figure 8-3:
Distribution
of the
lifetimes
of light
bulbs from
Lights Up.

Saddle point — 20 ← → 20 — Saddle point

940   960   980   1000   1020   1040   1060

Hours

Before examining any results, be sure to both examine the scale on the horizontal axis of any distribution and know what the standard deviation is. Depending on the scale used, a distribution can look more squeezed together or more spread out than it should. Figures 8-2 and 8-3, for example, look similar, but their scales are very different. A better way to compare the light bulb lifetimes of the two companies is to put them on the same scale, as shown in Figure 8-4. Now you can see how much more spread out the lifetimes are for the bulbs made by Lights Out compared to those made by Lights Up; the lifetimes of the bulbs made by Lights Up are much more concentrated around the mean.

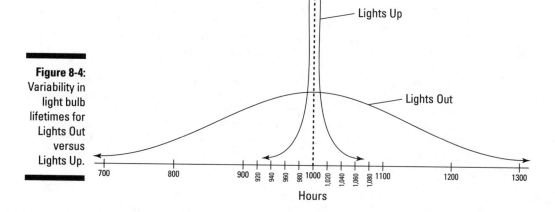

Figure 8-4:
Variability in
light bulb
lifetimes for
Lights Out
versus
Lights Up.

Lights Up

Lights Out

700   800   900   920   940   960   980   1000   1,020   1,040   1,060   1,080   1100   1200   1300

Hours

Looking for most of the values: The empirical rule

As long as a distribution has a mound shape in the middle — and the normal distribution certainly fits that criterion — you can make some general statements about where most of the values will be, using distances of 1, 2, or 3 standard deviations from the mean to mark off certain milestones. The rule that allows you to do this is called the *empirical rule.*

The empirical rule says that if a distribution has a mound shape, then:

✔ About 68% of the values lie within 1 standard deviation of the mean (or between the mean minus 1 times the standard deviation, and the mean plus 1 times the standard deviation). In statistical notation, this is represented as: $\mu \pm \sigma$.

✔ About 95% of the values lie within 2 standard deviations of the mean (or between the mean minus 2 times the standard deviation, and the mean plus 2 times the standard deviation). The statistical notation for this is: $\mu \pm 2\sigma$.

✔ About 99% (actually, 99.7%) of the values lie within 3 standard deviations of the mean (or between the mean minus 3 times the standard deviation and the mean plus 3 times the standard deviation). Statisticians use the following notation to represent this: $\mu \pm 3\sigma$.

In the formulas for the empirical rule, if you don't know the population mean and standard deviation, estimate (replace) the population standard deviation, σ, with the sample standard deviation, s. And you can also estimate (replace) the population mean, μ, with the sample mean, $\bar{x}$. See Chapter 3 for details.

Figure 8-5 illustrates the empirical rule. The reason that 68% of the values lie within 1 standard deviation of the mean is because the majority of the values on a normal distribution are mounded up in the middle, close to the mean (as Figure 8-5 shows). Remember, it has a bell shape. Moving out 1 more standard deviation on either side of the mean adds more values, but less than 30% more (for a total of 95% of the values) because now you're picking up less of the mound part and more of the tail part. Finally, going out 1 more standard deviation on either side of the mean gets you that last little bit of the tail areas, picking up 4.7% (nearly all of the rest) of the remaining values, to go from 95% to 99.7% of the data. Most researchers stay with the 95% range for reporting their results, because going out 3 standard deviations on either side of the mean doesn't seem worthwhile, just to pick up that last 4.7% of the values.

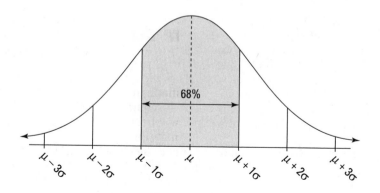

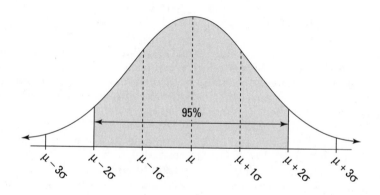

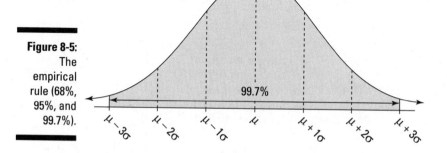

Figure 8-5:
The empirical rule (68%, 95%, and 99.7%).

I need to stress the word *about* in the preceding description of the empirical rule. These results are approximations only (but they're good approximations). Later in this chapter (see the "Converting to a Standard Score" section), you see how to give more precise information regarding what percent of the values in the distribution are between, above, or below certain values. However, the empirical rule is an important rule in statistics, and the concept of "going out two standard deviations gets you about 95% of the values" is one that you see mentioned often throughout this book.

With the light bulbs from Lights Out (refer to Figure 8-2), the standard deviation is 100 hours, and the mean is 1,000 hours. Using the empirical rule, you can discuss the relative standing of certain milestones in the data. For example, according to this model, about 68% of the light bulbs are expected to last between 900 and 1,100 hours (1,000 ± 100), about 95% of the light bulbs should last between 800 and 1,200 hours (1,000 ± 2 × 100), and 99.7% of the light bulbs should last between 700 and 1,300 hours.

You can use the symmetry of the normal distribution in combination with the empirical rule to answer other questions about the light bulb lifetimes. For example, what percentage of light bulbs should last 1,000 hours or more? The answer is 50%, because the median is at 1,000, and half of the values are greater than the median. What percentage of light bulbs from Lights Out should last more than 1,200 hours (refer to Figure 8-2)? The answer is 2.5%. Why? Because 95% of the light bulbs have lifetimes that are between 800 and 1,200 hours, and given that the total percentage under the whole curve has to be 100%, the remaining two tail areas must add up to 5%. Light bulbs lasting more than 1,200 hours make up the right tail only, and because of symmetry, you cut that 5% exactly in half to get 2.5%. So a light bulb that lasts more than 1,200 hours is pretty much a freak of nature, because that happens only 2.5% of the time (at least with Lights Out). With Lights Up (refer to Figure 8-3), a light bulb lasting that long would be unheard of, because 1,200 is much more than 3 standard deviations above the mean for the bulbs produced by that company (refer to Figure 8-4).

The moral of the story here is that if you like to gamble, buy your light bulbs from Lights Out, because you'll have a greater chance of getting either a very long-lasting bulb or a bulb that lasts for a very short time; in other words, the bulbs from Lights Out have more variability in their expected performance. If you're the conservative type, get your light bulbs from Lights Up; these bulbs are more consistent, with fewer surprises.

The empirical rule does *not* apply when a distribution doesn't have a mound shape in the middle. You can still approximate or determine where certain milestones are in the data by making a histogram and/or finding percentiles (see Chapters 4 and 5 for more on histograms and percentiles, respectively).

Converting to a Standard Score

Suppose that hypothetical physical therapy student Rhodie took a standard-ized test for certification to become a physical therapist, and her results indi-cated that she got a score of 235. All you know is that the scores for this test had a normal distribution. Is Rhodie's score a good one, a bad one, or is it just a middle-of-the-road result? You can't answer this question without a measure of where Rhodie stands among the other people who took the test. In other words, you need to determine the relative standing of Rhodie's test score on the distribution of all the scores.

Focusing on the standard deviation

You can determine the relative standing for Rhodie's test score in a number of ways — some better than others. First you can look at the score in terms of the total points possible, which for this test turns out to be 300. This doesn't compare her score to anyone else's, it just compares her score to the possible maximum. So you really don't know where Rhodie's score stands relative to the other scores. Next, you can try to compare her result to the average. Suppose the average was 250. This does provide a bit more information. At this point you know Rhodie's score of 235 is below average; in fact, her score is 15 points below average (because 235 − 250 = −15). But what does a differ-ence of 15 points mean in this situation?

As you see in the "Straghtening Out the Bell Curve" section earlier in this chapter, which discusses the light bulb lifetimes from the two companies (refer to Figure 8-4), to get an understanding of relative standing for any value on a distribution, you have to know what the standard deviation is. Suppose the standard deviation of the distribution of scores on Rhodie's test was 5 (as in Figure 8-6).

Figure 8-6:
Scores having a normal distribution with mean of 250 and standard deviation of 5.

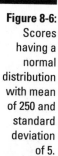
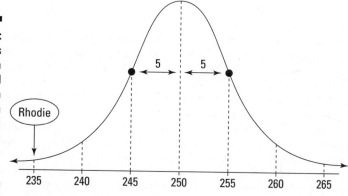

A distribution with a standard deviation of only 5 means that the scores were fairly close together, and 15 points below average is really quite a bit, relatively speaking. In this case, Rhodie's score is interpreted as being pretty far below average, because the 15 point difference is 3 standard deviations below the mean (because each standard deviation is worth 5, and $-15 \div 5 = -3$). Only a tiny fraction of the other test takers scored lower than she did. (You know that 99.7% of the test takers scored between 235 and 265 by the empirical rule, and the total percentage for all possible scores is 100. This means that the percentage who scored outside of the 235 to 265 range is $100 - 99.7 = 0.3\%$. You want the percentage falling below Rhodie's score of 235, which is half of 0.3%. This means that only 0.15%, or 0.0015 of the test takers scored lower than Rhodie in this case.)

Now suppose that the standard deviation is a different value, say 15, but the mean of the test scores is the same (250). Figure 8-7 shows what this distribution looks like.

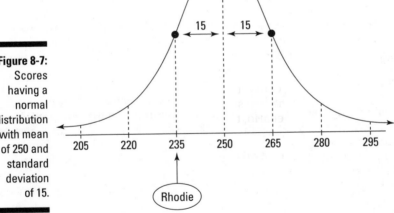

Figure 8-7:
Scores having a normal distribution with mean of 250 and standard deviation of 15.

A standard deviation of 15 means the scores are much more variable (or spread out) than they were in the previous situation. In this case, Rhodie's score being 15 points below average isn't so bad, because these 15 points represent only 1 standard deviation below the mean (because $-15 \div 15 = -1$). In this example, 68% of the scores are between 235 and 265 by the empirical rule, and half of the remaining 32% (those in the lower tail) scored lower than Rhodie. So in this scenario, 16% of the test takers scored lower than Rhodie. Her relative standing still isn't great, but this second scenario improves her relative standing compared to the first scenario. Notice that her score didn't change from one scenario to the other; what changed was the interpretation of her score due to the difference in the standard deviations.

The relative standing of any score on a distribution depends greatly on the standard deviation. Distances in original units don't mean much without it.

Many times, in the media, the standard deviation is never to be found. Never interpret any statistical result by comparing it only to the mean without knowing what the standard deviation is. Numbers may be farther from the mean than they appear.

Calculating the standard score

To find, report, and interpret the relative standing of any value on a normal distribution (such as Rhodie's score), you need to convert the score to what statisticians call a standard score. A *standard score* is a standardized version of the original score; it represents the number of standard deviations above or below the mean. The formula for calculating a standard score is the following:

Standard score = (original score − mean) ÷ standard deviation. Or, using shorthand notation, standard score = $\frac{x - \mu}{\sigma}$.

To convert an original score to a standard score:

1. **Find the mean and the standard deviation of the population that you're working with.**

   For example, you can convert Rhodie's exam score of 235 to a standard score under each of the two scenarios discussed in the preceding section. In the first scenario, the mean is 250 and the standard deviation is 5, so Step 1 is done.

2. **Take the original score, and subtract the mean.**

   In Rhodie's first scenario, you find the actual distance from the mean by taking 235 − 250 = −15 (which means her score is fifteen points below the mean).

3. **Divide your result by the standard deviation.**

   In Rhodie's case, the distance is −15. Converting this distance in terms of number of standard deviations means taking −15 ÷ 5 = −3, which is Rhodie's standard score. In the first scenario (standard deviation = 5), Rhodie's score of 235 is 3 standard deviations below the mean.

   In the second scenario (standard deviation = 15), Rhodie's standard score is (235 − 250) ÷ 15 = −15 ÷ 15 = −1. So her score is 1 standard deviation below the mean in the second scenario.

To avoid errors in converting to standard scores, be sure to do Steps 2 and 3 in the order given.

Properties of standard scores

The following properties may prove helpful when interpreting standard scores:

- ✔ Almost all standard scores (99.7% of them) fall between the values of –3 and +3, because of the empirical rule.

- ✔ A negative standard score means the original score was below the mean.

- ✔ A positive standard score means the original score was above the mean.

- ✔ A standard score of 0 means the original score was the mean itself.

- ✔ Scores that come from a normal distribution, when standardized, have a special normal distribution with mean 0 and standard deviation 1. This distribution is called the *standard normal distribution* (see Figure 8-8).

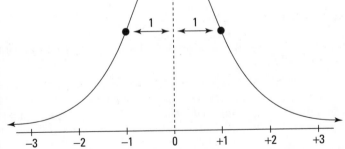

Figure 8-8:
The standard normal distribution.

Standard scores have a universal interpretation, which is what makes them so great. If someone gives you a standard score, you can interpret it right away. For example, a standard score of +2 says that this score is 2 standard deviations above the mean. To interpret a standard score, you don't need to know what the original score was or what the mean and standard deviation were. The standard score gives you the relative standing of that value, which, in most cases, is what matters most.

Converting to standard scores doesn't change the relative standing of any of the values on the distribution; it simply changes the units. (This is analogous to changing the units of temperature when converting from the Fahrenheit scale to the Celsius scale. The temperature outside doesn't change, but the units used to measure that temperature do.) Subtracting the mean from the original score centers everything at 0. If the original score falls directly on the mean, it will be converted to a standard score of 0. The tick marks on either side of the mean are still each in terms of the original standard deviation (which can go by 5s, 15s, and so on). For standard scores, you want them to be in units of 1, instead, for ease of interpretation. That means after you subtract the mean, you must divide your result by the standard deviation. (This is similar to changing from inches to feet by dividing by 12.)

Comparing apples and oranges using standard scores

One common use for standard scores is to compare scores from different distributions that would otherwise not be comparable. For example, suppose Bill applies to two different colleges (call them Data University and Ohio Stat), and he has to take a math placement test for each of the colleges. The tests are totally different (they even had a different number of questions) and when Bob gets his scores back, he wants to be able to compare his scores and determine which college gives him a better relative standing on its math placement test.

Data University tells Bob that his score is 60 and that the distribution of all scores is normal with a mean of 50 and standard deviation of 5. Ohio Stat tells Bob that his score is 90 and that the scores have a normal distribution with mean 80 and standard deviation 10. On which test does Bob perform better? You can't compare the 50 to the 90 outright, because they're on totally different scales. And you can't say that he performed the same on each just because he was 10 points above the average on each one. Here, the standard deviation is an important factor. The way to make a fair and equitable comparison of these scores is to convert each of them to standard scores, allowing them to both be on the same scale (where most of the values lie between −3 and +3 with units of 1).

Again, Bob's score on Data University's placement exam is 60 with a mean of 50 and standard deviation of 5 for the population of all test scores. His standard score, therefore, is $(60 - 50) \div 5 = 10 \div 5 = +2$, meaning his relative standing on Data University's placement exam is 2 standard deviations above the mean. His score on Ohio Stat's placement test is 90, where the mean is 80 and the standard deviation is 10. His standard score, therefore, is $(90 - 80) \div 10 = 10 \div 10 = +1$, meaning Bob's relative standing on Ohio Stat's math placement exam is 1 standard deviation above the mean. This score is not as high, relatively speaking, as his Data University math placement exam score. Bob, therefore, performs better on Data University's math placement exam.

Don't compare results from different distributions without first converting everything to standard scores. Standard scores allow for fair and equitable comparisons on the same scale.

Sizing Up Results Using Percentiles

Percentiles are used in a variety of ways for comparison purposes and to determine relative standing. Babies' weights, lengths, and head circumferences are reported and interpreted in terms of percentiles, for example. Percentiles are also used by companies to get a handle on where they stand in terms of the competition on sales, profits, customer satisfaction, and so on. (For the details on percentiles, see Chapter 5.) And the relationship between percentiles and standard scores is an important one, as the next example shows.

Suppose Rhodie (a hypothetical physical therapy student) takes a physical therapy certification test and gets a score of 235. She's trying to figure out what that score really means. She is told that the test scores have a normal distribution with mean 250 and standard deviation 15, which means her standard score on the test is −1.0 (one standard deviation below the mean). Her score is below average, but is it still good enough to pass the test? Each year, the certification test is different, so the cutoff score for passing and failing changes, too. However, the test administrators always pass the top 60% of the scores and fail the bottom 40%. Knowing this information, does Rhodie pass the test? And what's the cutoff score for this year's test? Percentiles help Rhodie unravel the mystery of her standard score and relieve her anxiety.

If the test administrators always pass the top 60% and fail the bottom 40%, that means the cutoff score for pass/fail is at the 40th percentile. (Remember that percentile means percentage *below*.) At what percentile is Rhodie's score? Referring to Figure 8-7 and using the empirical rule, you know that her score is 1 standard deviation below the mean. So, about 68% of the scores lie between 235 and 265 (within 1 standard deviation of the mean) and the rest (100% − 68% = 32%) lie outside that range. Half of those outside the ±2 standard deviation range (or 32% ÷ 2 = 16%) lie below 235. This means that Rhodie's score is at the 16th percentile (approximately), so she doesn't pass the test. To pass, she needs to at least be at the 40th percentile or better.

The empirical rule can take you only so far in determining percentiles; if you notice, Rhodie's score was conveniently chosen by me so that it fell right on one of the tick marks. But suppose she scored somewhere in between the tick marks? Never fear, the standard normal table is here. Table 8-1 shows the corresponding *percentile* (percentage below) for any standard score between −3.4 and +3.4. This covers well over 99.7% of the situations you'll ever come across. Notice that as the standard scores get larger, the percentile also gets larger. Also note that a standard score of 0 is at the 50th percentile point of the data, which is the same as the median. (See Chapter 5 for more on means and medians.)

Table 8-1		Standard Scores and Corresponding Percentiles from the Standard Normal Distribution			
Standard Score	Percentile	Standard Score	Percentile	Standard Score	Percentile
−3.4	0.03%	−1.1	13.57%	+1.2	88.49%
−3.3	0.05%	−1.0	15.87%	+1.3	90.32%
−3.2	0.07%	−0.9	18.41%	+1.4	91.92%
−3.1	0.10%	−0.8	21.19%	+1.5	93.32%
−3.0	0.13%	−0.7	24.20%	+1.6	94.52%
−2.9	0.19%	−0.6	27.42%	+1.7	95.54%
−2.8	0.26%	−0.5	30.85%	+1.8	96.41%
−2.7	0.35%	−0.4	34.46%	+1.9	97.13%
−2.6	0.47%	−0.3	38.21%	+2.0	97.73%
−2.5	0.62%	−0.2	42.07%	+2.1	98.21%
−2.4	0.82%	−0.1	46.02%	+2.2	98.61%
−2.3	1.07%	0.0	50.00%	+2.3	98.93%
−2.2	1.39%	+0.1	53.98%	+2.4	99.18%
−2.1	1.79%	+0.2	57.93%	+2.5	99.38%
−2.0	2.27%	+0.3	61.79%	+2.6	99.53%
−1.9	2.87%	+0.4	65.54%	+2.7	99.65%
−1.8	3.59%	+0.5	69.15%	+2.8	99.74%
−1.7	4.46%	+0.6	72.58%	+2.9	99.81%
−1.6	5.48%	+0.7	75.80%	+3.0	99.87%
−1.5	6.68%	+0.8	78.81%	+3.1	99.90%
−1.4	8.08%	+0.9	81.59%	+3.2	99.93%
−1.3	9.68%	+1.0	84.13%	+3.3	99.95%
−1.2	11.51%	+1.1	86.43%	+3.4	99.97%

In order to use Table 8-1 to find a percentile, you must first convert the original score to a standard score. That's easier than carrying around a table for every possible normal distribution with every possible mean and standard deviation, right? That's why the standard normal distribution is so great; it uses a standardized scale for one-table-fits-all usage.

To calculate a percentile when the data have a normal distribution:

1. **Convert the original score to a standard score by taking the original score minus the mean, and then dividing that difference by the standard deviation. (The notation for this is $\frac{x-\mu}{\sigma}$.)**

2. **Use Table 8-1 and find the corresponding percentile for the standard score.**

You already know that Rhodie scored at the 16th percentile with her score of 235. Suppose Clint also takes the same test, and he scores 260. Does he pass? To find out, you can convert his score to a standard score and find the corresponding percentile. His standard score is $(260 - 250) \div 15 = 10 \div 15 = 0.67$. Using Table 8-1, Clint's score is somewhere between the 72.58th percentile and the 75.80th percentile. (To be conservative, go with the lower percentile.) In any case, Clint does pass, because he performs better than at least 72% of the test takers (including Rhodie). And his percentile is higher than 40, beating the cutoff percentile.

Wouldn't it be nice not to have to convert every person's score to a standard score just to determine whether he or she passed? Why not just find the cutoff score, in terms of original units, and compare everyone's result to that? You know that the cutoff score is at the 40th percentile. Using Table 8-1, you can see that the corresponding standard score closest to the 40th percentile is –0.3 (use the table backward). This means that the cutoff score is 0.3 standard deviations below the mean. What's this score in standard units? You can use the standard score formula and solve it backwards to find the original score.

Recall that the formula for converting an original score (x) to a standard score (call it Z) is $Z = \frac{x-\mu}{\sigma}$. With a little algebra, you can rewrite this equation so that you can convert from a standard score (Z) back to the original score (x). That formula looks like this: $x = Z\sigma + \mu$.

To convert a standard score to a score in original units (original score):

1. **Find the mean and the standard deviation of the population that you're working with.**

2. **Take the standard score (Z) and multiply by the standard deviation.**

3. **Add the mean to your result from Step 2.**

Going back to the example, you know that the pass/fail cutoff as a standard score is –0.3, so Z = –0.3. You also know that the mean of all the scores is μ = 250, and the standard deviation is σ = 15. Converting the standard score to the original score, you get $x = -0.3 \times 15 + 250 = -4.5 + 250 = 245.5$ (or 246). So the cutoff score for pass/fail is 246. Anyone scoring below 246 fails (Rhodie, for example) and anyone scoring at or above 246 passes (Clint, for example).

What if the scores don't have a normal distribution? You can still calculate percentiles, but you will have to do it manually, or use some computer software to do it (such as Microsoft Excel).

To find the kth percentile when the data do not have a normal distribution:

1. **Put the values in order from smallest to largest.**

2. **Let *n* be the size of the data set. Multiply *k* percent times *n*, and round to the nearest whole number.**

3. **Count your way through the data until you reach the point identified in Step 2. This is the *k*th percentile in your data set.**

For example, suppose you have the following data set: 1, 6, 2, 5, 3, 9, 3, 5, 4, 5, and you want the 90th percentile.

Step 1: order the data to get 1, 2, 3, 3, 4, 5, 5, 5, 6, 9.

Step 2: $n = 10$, $k = 90\%$, and k percent times n is 0.90 times 10 = 9.

Step 3: That means the 9th number (from smallest to largest), which is 6, is the 90th percentile. About 90% of the values are below 6, and 10% of the values are at or above 6. (See Chapter 5 for more discussion and examples of percentiles.)

Chapter 9

Caution: Sample Results Vary!

· ·

In This Chapter

▶ Realizing that sample results vary

▶ Measuring variability in sample results with the central limit theorem

▶ Determining the factors that affect variability

· ·

Statistics are often presented as one-shot deals. For example, "One out of every two marriages ends in divorce," "Four out of five dentists surveyed recommend Trident gum," or "The average lifespan of a female born in the year 2000 is 80 years." People hear statistics like this and assume that those results apply to them. The ordinary person may assume, for example, that his chance of getting divorced is 50%, that his dentist probably recommends Trident gum, and that if he and his wife (if they haven't divorced yet!) had a baby girl in the year 2000, they can expect her to live to be 80 years old.

But shouldn't these statistics come with a "plus or minus" indicating that the results vary? You bet! Does it happen? Not often enough. The truth is that unless the researchers are able to conduct a *census* to get their results (collecting data on every single member of the population), those results are going to vary from sample to sample, and that variability can be much more than you think! The question is, by how much should you expect a statistical result to vary? You hope (perhaps even automatically assume) that it shouldn't vary by much, and that you can accurately apply the reported result to almost anyone. But is this always the case? Absolutely not; the variability in any statistical result depends on a number of factors, all of which are discussed in this chapter.

Expecting Sample Results to Vary

I was watching a commercial on TV the other day for a weight-loss meal-replacement drink. It gave an inspiring story about a woman who had lost 50 pounds in 6 months (and who had kept off the weight for over a year). During her testimony, a message flashed for a couple of seconds at the bottom of the screen that said "results not typical."

That leads to the question, "What's typical?" How much weight can you expect to lose in 6 months on this product, or, if you wanted to lose 50 pounds, how long should you expect it to take? You know that no matter what the results are for any individual, results are expected to vary from person to person. But this commercial is trying to lead you to believe that you should expect to lose about 50 pounds in 6 months (even though the tiny message says you shouldn't). What would be nice is if this manufacturer told you by how much you should expect those results to vary. What would also be nice is if the commercial presented the results from a sample of people, not just from one person.

Anecdotes (individual stories and testimonies) are eye catching, but they're not statistically meaningful!

Suppose you're trying to estimate the proportion of people in the United States who approve of the president. If you ask a random sample of 1,000 people from the United States whether they approve of how the president is doing his job, you'll get one sample result (for example, 55% approve). You shouldn't report that 55% of the entire population of the United States approves of the president, because your result is based on one sample of only 1,000 people.

If you take a different random sample of 1,000 people from the same population and ask the same question, you'll likely get a different result. In fact, given that the population of the United States is so large and not everyone shares the same opinion of the president, 100 different random samples of 1,000 people — each taken from the same population and each asked the same question — would yield 100 different results. So how do you report your sample results? Some measure of how much the results are expected to vary has to be part of the package.

Expect sample results to vary from sample to sample. Don't take a statistic at face value and try to apply it without having some indication of how much that result is expected to vary.

Measuring Variability in Sample Results

You may be wondering how you can assess the amount by which a sample statistic is going to vary without having to select every possible sample and look at its statistic. You may as well do a census at that point, right? Fortunately, thanks to some major statistical results (mainly the central limit theorem), you can find out how much you expect sample means or proportions to vary without having to take all possible samples (what a relief!). The *central limit theorem,* in a nutshell, says that the distribution of all sample means (or proportions) is normal, as long as the sample sizes are large enough. And even more impressive, the central limit theorem doesn't care what the distribution of the original population looks like. How can this be? You need to take a large enough sample, and you need to know some of the characteristics of your original population (like the mean and the standard deviation). And then the magic of statistical theory takes over from there.

Standard errors

Variability in sample means (or proportions) is measured in terms of standard errors. *Standard error* is the same basic concept as standard deviation; both represent a typical distance from the mean. But here is how standard error differs from standard deviation. The original population values *deviate* from each other due to natural phenomena (people have different heights, weights, and so on), hence the name standard *deviation* to measure their variability. Sample means vary because of the error that occurs in not doing a census and being able to only take samples, hence the name standard *error* to measure the variability of the sample means. (See Chapter 5 for more on the standard deviation. I talk more about how to interpret standard errors in this section. For specifics on standard error formulas, see Chapter 10.)

Here's an example: The U.S. Bureau of Labor Statistics tries to track what people spend their money on each year with a Consumer Expenditure Survey (CES). The bureau takes a sample of households and asks each household in the sample to give their spending information. (Bias in reporting could be an issue here.) Their typical sample size is 7,500 people. Table 9-1 shows a few of the results from the 2001 CES. This table not only includes the average amount of money spent by people in the sample on various items (the sample means for each item), it also includes the standard error for each of those sample means.

Table 9-1	Average Yearly Household Expenses for American Households in 2001	
Expense	*Mean*	*Standard Error*
Food (eating at home)	$3,085.52	$42.30
Food (eating out)	$2,235.37	$38.35
Phone	$914.41	$9.69
Gas and oil (for vehicles)	$1,279.37	$12.88
Reading materials	$141.00	$2.99

You can interpret the results of Table 9-1 by making relative comparisons. For example, notice that about 42% of all average household food expenses are for eating out, given that $2,235.37 ÷ ($3,085.52 + $2,235.37) = $2,235.37 ÷ $5,320.89 = 0.42 or 42%. The standard errors for average food expenses are larger than for the other expenses on the list, because food expenses vary a lot more from household to household. However, you may wonder why the standard errors for food expenses aren't larger than what's shown in Table 9-1. Remember, standard error tells you how much variability you can expect in the average if you were to take another sample. If the sample size is large, the average shouldn't change by much. And you know that the government never uses small sample sizes!

A listing of the standard errors for sample means is not something you would typically see in a media report. However, you can (and should, when the results are important to you) dig deeper and find the standard errors, as well. The best thing to do is to look up the research articles and look for the standard errors in those articles.

Sampling distributions

A listing of all the values that a sample mean can take on and how often those values can occur is called the *sampling distribution* of the sample mean. A sampling distribution, like any other distribution, has a shape, a center, and a measure of variability (in this case, the standard error). (See Chapter 4 for information on shape, center, and variability; see Chapter 3 for more information on distributions.)

According to the central limit theorem, if the samples are large enough, the distribution of all possible sample means will have a bell-shaped, or normal distribution with the same mean as that of the original population. (See Chapter 3 for more on the normal distribution.) This is because the sample means are clustered near the overall average value, which is the population mean. High values in a sample are offset by low values that also appear in the sample, in an *averaging out* effect. The variability in the sampling distribution is measured in terms of standard errors. An added benefit to using an average to get an estimate (rather than a total or a single value) is that the variability in the sample means decreases as sample sizes get larger. (Similar characteristics also apply to the sampling distribution of the sample proportion, in the case of categorical [yes/no] data from surveys and polls.)

Using the empirical rule to interpret standard errors

Because the sampling distribution of sample means (or sample proportions) is normal (mound-shaped), you can use the empirical rule to get an idea of how much a given sample result is expected to vary, provided that the sample size is large enough. (See Chapter 8 for full coverage of the empirical rule.)

Applied to sample means and proportions, the empirical rule says that you can expect:

- ✔ About 68% of the sample means to lie within 1 standard error of the population mean

- ✔ About 95% of the sample means to lie within 2 standard errors of the population mean

- ✔ About 99.7% of the sample means to lie within 3 standard errors of the population mean

- ✔ Similar values for categorical (yes/no) data: 68%, 95%, or 99.7% of the sample proportions will lie within 1, 2, or 3 standard errors, respectively, of the population proportion

What does the empirical rule tell you about how much you can expect a given sample mean to vary? Keep in mind that 95% of the sample means should lie within 2 standard errors of the population mean, and your job is to estimate the population mean. So, if your estimate is actually a range including your sample mean plus or minus 2 standard errors, your estimate would be correct about 95% of the time. (The number of standard errors added or subtracted is called the *margin of error.* For more on the margin of error, see Chapter 10.)

Consider this example: According to the U.S. Bureau of Labor Statistics, the average household in 2001 contained 2.5 people (0.7 of which were children under 18, and 0.3 of which were people over 65) and 1.9 vehicles. (Sorry, no standard errors were available for these data.) Referring to Table 9-1, the average phone expenses for the year for this sample of 7,500 households was $914.41 per household. How much are these results expected to vary from sample to sample (that is, if different samples of 7,500 households each had been selected from the same population)? The standard error for phone expenses for this sample is $9.69. This means that 95% of the sample average phone expenses should lie within 2 × $9.69 (or $19.38) on either side of the actual population average. This shows how much the mean phone expenses are expected to vary when the sample size is 7,500.

In the preceding example, you're not saying that 95% of all the households in the population have phone expenses in that range. Instead, you're giving an estimate for what the average phone expense is, over all households in the population. The average phone expense would actually be a single number. But because you can't get the actual number, you estimate it using this range of values.

You can also use the empirical rule to give a rough estimate of the average telephone expenses for all U.S. households (not just the sample of 7,500 people). Again, using the second property of the empirical rule, you would expect that the average telephone expense for all households in the U.S. is about $914.41, plus or minus $2 \times \$9.69 = \$19.38$. This type of estimate will be correct for 95% of the samples selected (and you hope that the sample collected in the Consumer Expenditure Survey is one of them). Using proper statistical jargon, this means that you can estimate with about 95% confidence that the average phone expenses per year for all U.S. households lies somewhere between $914.41 – $19.38 and $914.41 + $19.38, or between $895.03 and $933.79. If you want to be about 99.7% confident in your estimate (instead of 95% confident), you need to add and subtract 3 standard errors.

This type of result, involving a statistic plus or minus a certain number of standard errors, is called a *confidence interval*. (For more on this see Chapter 11.) The amount added or subtracted is called the *margin of error*.

Reporting the margin of error for the sample mean is something that you don't see very often in media reports involving quantitative data (such as household income, house prices, or stock market values). Yet the margin of error should always be there in order for the public to assess the accuracy of the results! With survey and polling data (which are categorical data and are reported as proportions) you're often given the margin of error, which is directly related to the standard error (see Chapter 10). Why the double standard? I can't say for sure.

Specifics of the central limit theorem

Notice that the results that apply the empirical rule in the preceding section give a rough way of interpreting 1, 2, or 3 standard errors, and they tell you what to expect in terms of the sample mean (or sample proportion). These results are actually due to the central limit theorem (CLT). Statisticians love the CLT; without it, they wouldn't have jobs. The CLT allows them to actually say something about where they expect sample results to lie, without having to look at all possible samples drawn from a particular population. It also provides a formula for calculating standard errors, as well as more specific information regarding what percentage of the sample means (or proportions) will lie between any number of standard errors (not just 1, 2, or 3).

The central limit theorem says that for any population with mean μ and standard deviation σ:

> ✔ The distribution of all possible sample means, $\bar{x}$, is *approximately* normal for sufficiently large sample sizes. That means you can use the normal distribution to answer questions or to draw conclusions about your sample mean. (See Chapter 8 for the normal distribution.)

✔ The larger the sample size (*n*) is, the closer the distribution of the sample means will be to a normal distribution. (Most statisticians agree that if *n* is at least 30, it will do a reasonable job in most cases.)

✔ The mean of the distribution of sample means is also μ.

✔ The standard error of the sample means is $\frac{\sigma}{\sqrt{n}}$. It decreases as *n* increases.

✔ If the original data have a normal distribution, the sample means will always have an exact normal distribution, no matter what the sample size is.

If the population standard deviation, σ, is unknown (which will be the case most of the time), you can estimate it using *s*, the sample standard deviation, for standard error in the preceding formula. More on this in Chapter 12.

Note that the CLT says that even if the original data are not normally distributed, the distribution of sample means will be normal, as long as the sample sizes are large enough. This, again, is due to the averaging effect.

Checking out the ACT math scores

Consider the 2002 ACT math scores for male and female high school students. (This is a situation in which the population mean and standard deviation are known, because all of the tests taken in 2002 were graded and recorded.) The average ACT math score for male students was 21.2 with a standard deviation of 5.3. Female students averaged 20.1 (with a standard deviation of 4.8) on the same ACT math test. Prior research has shown that ACT scores have an approximate normal distribution. Figure 9-1 shows the distributions of the scores for males and females, respectively, given the preceding information. In each case, the size of the entire test-taking population (boys and girls combined) is about one million students.

Using the empirical rule (see Chapter 8), about 95% of the male students scored between 10.6 and 31.8 on the ACT math test, and about 95% of the female students scored between 10.5 and 29.7 on the ACT math test. The scores of male and female students are quite comparable.

Suppose that you're interested in the average scores of samples of size 100 from the total population of 500,000 male students who took the test in 2002. Why? Maybe you have 100 students in a class and want to see how they did, compared to all other possible classes of this size. What will the distribution of all possible sample means look like? According to the CLT, it will have a normal distribution with the same mean (21.2), and the standard error will be 5.3 divided by the square root of 100 (because in this hypothetical sample, *n* = 100). Therefore the standard error for this sample is $\frac{5.3}{\sqrt{100}}$ = 5.3 ÷ 10 = 0.53. Figure 9-2 shows what the sampling distribution of the sample means looks like for samples of 100 male students.

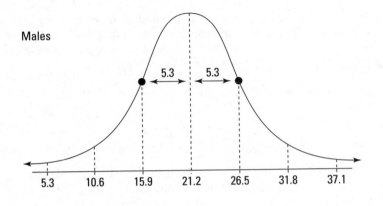

Males

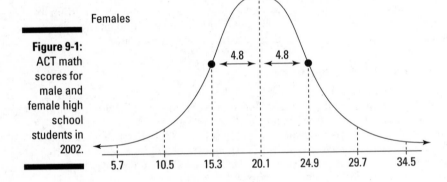

Females

Figure 9-1:
ACT math
scores for
male and
female high
school
students in
2002.

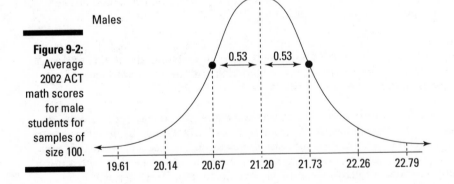

Males

Figure 9-2:
Average
2002 ACT
math scores
for male
students for
samples of
size 100.

Notice in Figure 9-2 how much smaller the standard error of the sample means is, compared to the standard deviation of the original scores shown in Figure 9-1. That's because each sample mean in Figure 9-2 contains information from 100 students, compared to each individual score in Figure 9-1, which contains information from only a single student. Sample means won't vary as much as individual scores will. That's why using a sample mean is a much better idea for estimating the population mean than just using an individual score (or an anecdote).

Figure 9-3 shows what happens to the sampling distribution of the sample mean when the sample sizes increase to 1,000 male students. The standard error reduces to 5.3 divided by the square root of 1,000, or 5.3 ÷ 31.62 = 0.17. The standard error for Figure 9-3 is smaller than the standard error in Figure 9-2, because the sample means in Figure 9-3 are each based on 1,000 students and contain even more information than the sample means shown in Figure 9-2, (which are based on 100 students each).

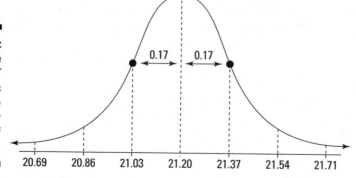

Figure 9-3: Average 2002 ACT math scores for male students for samples of size 1,000.

0.17   0.17

20.69   20.86   21.03   21.20   21.37   21.54   21.71

Figure 9-4 shows all three distributions for male students (individual scores, sample means of size 100, and sample means of size 1,000) overlapping, to compare their variability.

You can use the central limit theorem to answer questions about sample results in situations like the ACT math scores. For example, suppose you want to know the chance that a sample of 100 male students will have an average ACT math score of 22 or less. Using the technique shown in Chapter 8, you change the score of 22 to a standard score by subtracting the population mean (21.2) and dividing the difference by the standard error (instead of the standard deviation). The formula for this conversion is $\dfrac{\bar{x} - \mu}{\sigma \div \sqrt{n}}$, where $\bar{x}$ is the average score of the sample (in this case, 22) and $\dfrac{\sigma}{\sqrt{n}}$ is the standard error. Note that σ is the population standard deviation (5.3). In this example, where the standard error was calculated to be $5.3 \div \sqrt{100} = 0.53$ (see Figure 9-2), the class average of 22 converts to a standard score of $(22 - 21.2) \div 0.53 = 0.8 \div 0.53 = 1.51$. You want to know the percentage of scores that lie to the left of this value (in other words, the percentile corresponding to a standard score of 1.51). Referring to Table 8-1 in Chapter 8, that percentage is about 93.32%.

Don't forget to find $22 - 21.2$ first, before dividing by 0.53 in the precdeing example, or you'll get –18, which is wrong.

A group of averages always has less variability than a group of individual scores. And averages that are based on larger sample sizes have even less variability than averages based on smaller sample sizes.

The central limit theorem doesn't apply only to sample means. You can also use the central limit theorem to answer questions or make conclusions about population proportions, based on your sample proportion. The same conclusions about the shape, center, and variability in the sample means applies to sample proportions. Of course, the formulas will be a little different, but the concepts are all the same. First, the sample proportion is denoted $\hat{p}$ and is equal to the number of individuals in the sample in the category of interest, divided by the total sample size (n).

Figure 9-4:
Sampling
distributions
of 2002 ACT
math scores
for male
students
showing
original
scores,
sample
means for
samples of
size 100, and
sample
means for
samples of
size 1,000.

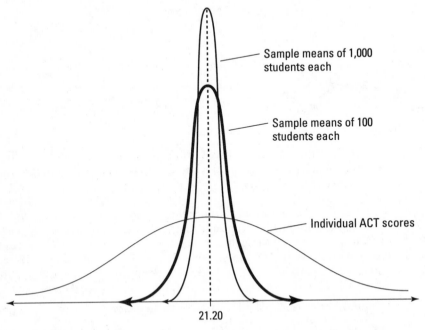

The central limit theorem says that for any population of data with p as the overall population percentage:

> ✔ The distribution of all possible sample proportions $(\hat{p})$ is *approximately* normal, provided that the sample size is large enough. (See Chapter 8 for more on the normal distribution.)

> ✔ The larger the sample size (n) is, the closer the distribution of sample proportions will be to a normal distribution. (That means you can use the normal distribution to answer questions or to draw conclusions about your sample proportion.)

> ✔ The mean of the distribution of sample proportions is also p.

> ✔ The standard error of the sample proportions is $\sqrt{\dfrac{p(1-p)}{n}}$. It decreases as n increases.

Notice that the standard error of the sample proportion actually contains p, which is the population proportion. That value will most likely be unknown; you can estimate it with the sample proportion, $\hat{p}$. More on this in Chapter 12.

What proportion needs math help?

You can use the central limit theorem to answer questions involving proportions. For example, suppose you want to know what proportion of incoming college students would like some help in math. A student survey accompanies the ACT test each year, and one of the questions asked is whether each student would like some help with his or her math skills. In 2002, 38% of the students taking the ACT test responded yes to this question. This is a situation in which the population proportion, p, is known ($p = 0.38$). The original data in this case (as with all categorical data) do not have a normal distribution because only two results are possible: yes or no. The distribution of the population of answers to the math skills question is shown in Figure 9-5 as a bar graph (see Chapter 4 for more information on bar graphs).

Suppose you were to take samples of size 100 from this combined population of over a million students (all of the students who took the ACT test in 2002), and find the proportion who indicated that they needed help with their math skills in each case. The distribution of all sample proportions is shown in Figure 9-6. It is a normal distribution with mean $p = 0.38$ and standard error = $\sqrt{\dfrac{0.38(1 - 0.38)}{100}} = \sqrt{0.00236} = 0.049$ or 4.9%, or about 5%. Using the CLT, you can say that some of the sample proportions are higher than 0.38, some are lower, but most of them (about 95% of them) lie in the area of 0.38 plus or minus $2 \times 0.05 = 0.10$, or 38% ± 10%. These results still vary by quite a bit, by 10% on either side of the population proportion.

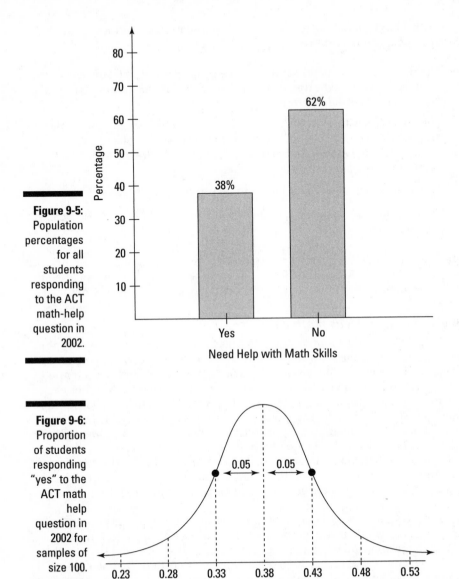

Figure 9-5:
Population
percentages
for all
students
responding
to the ACT
math-help
question in
2002.

Figure 9-6:
Proportion
of students
responding
"yes" to the
ACT math
help
question in
2002 for
samples of
size 100.

Now take samples of size 1,000 from the original population and find the proportion who responded that they needed help with math skills for each sample. The distribution of sample proportions in this case will look much like Figure 9-7. Everything will look the same as Figure 9-6, except that the distribution will be tighter; the standard error would reduce to $\sqrt{\frac{0.38(1-0.38)}{1,000}} = \sqrt{0.000236} = 0.015$, or 1.5%. About 95% of the sample results will lie between $0.38 - 2(0.015)$ and $0.38 + 2(0.015)$, or between 0.35 and 0.41 (that is, between 35% and 41%). In other words, if you take several different samples all of size 1,000 from this population and find the sample proportion for each sample, your sample proportions won't change much from sample to sample. Instead, they'd all be quite close together: That's due to the high sample size of 1,000.

Figure 9-7:
Proportion of students responding "yes" to the ACT math help question in 2002 for samples of size 1,000.

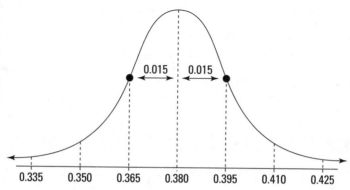

Before you draw conclusions from any sample percentages, get some idea of how much the results should vary by finding the standard error or the margin of error (which is about two standard errors; see Chapter 10). Knowing the expected amount of variability will help you keep the results in perspective.

How large is large enough for the central limit theorem to work for categorical data? Most statisticians agree that you should have $n \times p$ and $n \times (1-p)$ both be greater than or equal to 5. This takes care of any situations in which the proportion is very close to either 1 or 0 (in other words, those extreme situations where either almost everybody or almost nobody is in the group of interest). In these extreme situations, you'd need a larger sample to ensure that all the groups are represented, even those that don't contain many people. Most surveys and polls easily sample enough people to take care of this condition.

The CLT is good news for people who are trying to interpret sample results. As long as the sample size is large enough (and the data are credible and unbiased), the information reported will be close to the truth. (But remember, I said as long as the results are credible and unbiased. See Chapter 2 for examples of how statistics can go wrong.)

The central limit theorem also allows you to answer other important questions regarding sample means and proportions. For example, if a package delivery promises an average delivery time of two days, and your sample of 30 packages took 2.4 days, is this enough evidence to say that the company is guilty of false advertising? Or was this just an atypical sample of late packages? I address this type of question in Chapter 14.

If you're worried that you always need to know the population mean (μ) or the population proportion (p) in order to use the CLT, never fear! You will find out the secret that statisticians have known for years: If you don't know what a certain value is, just estimate it and move on. (More on this in Chapter 11.)

Examining Factors That Influence Variability in Sample Results

Two major factors influence the amount of variability in a sample mean or sample proportion: the size of the sample and the amount of variability in the original population.

Sample size

The size of the sample affects the amount of variability in the sample results. Suppose that you have a pond of fish, and you want to find the average length of all the fish in the pond. If you take repeated random samples of size 100 and repeated random samples of size 1,000, recording the sample mean each time, which sample means would vary more, those of size 100 or those of size 1,000? Those of size 100 would vary more, because each of the sample means was based on less information (that is, on fewer fish). Sample proportions would be affected similarly.

Small sample sizes result in sample means (and sample proportions) with large standard errors. Larger sample sizes result in sample means (and sample proportions) with smaller standard errors. In other words, the more data you collect with a single sample, the less variability you should have from sample to sample.

Variability in the sample means (or in the sample proportions) is measured by the standard errors. The variability of the sample means is $\frac{\sigma}{\sqrt{n}}$, and the variability in the sample proportions is $\sqrt{\frac{p(1-p)}{n}}$. The denominator of each of these formulas has n in it (and nothing else). Therefore, as the sample size (the denominator) increases, the standard error (the entire fraction) decreases. More information provided by the sample (through larger sample sizes) decreases the variability in the sample means (and in the sample proportions).

Population variability

As the variability in the population increases, so does the variability in the sample mean or sample proportion. Suppose that you have two ponds of fish, and you want to find the average length of all the fish in each pond. The fish in Pond Vary-Lot are much more variable in length than the fish in Pond Vary-Little are. You take a sample of 100 fish from each pond and find the mean length of the fish in your sample. If you take repeated samples of size 100 from each pond and record the sample mean in each case, which sample means will vary more, those from Pond Vary-Lot or those from Pond Vary-Little? The sample means from Pond Vary-Lot would vary more, because the population of fish in Pond Vary-Lot were more variable in their lengths to begin with.

Variability in sample proportions is affected in a similar way by the variability in the population. For example, suppose you want to estimate the proportion of fish in Pond Vary-Little that are in good health (call it p). If the fish in Pond Vary-Little are almost all either in good health (meaning p is close to 1), the standard deviation of the population, $p(1-p)$, is going to be small because most of the fish have the same health status. If you then take many samples of fish from this homogeneous (health-wise) population and find the percentage that is in good health, you shouldn't expect that percentage to change much from one sample to the next. So the standard error of the sample proportion is small when p is close to 1. The same thing happens when most of the fish are in poor health (p is close to 0). However, if about 50% of the fish are in good health and 50% are in poor health, you'll see more variability in your sample proportions from sample to sample, because the population has more variability in its health. In fact, a population where p is equal to 0.5 has the most variability in it, resulting in the standard errors of the sample proportions to be at their largest, as well.

More variability in the original population contributes more variability to the standard error of the sample means (or the sample proportions). Note that this increased variability can be offset, however, by increasing the sample size, as discussed above.

Recall that the variability of the sample means is $\frac{\sigma}{\sqrt{n}}$, and the variability in the sample proportions is $\sqrt{\frac{p(1-p)}{n}}$. The numerator of each of these formulas is actually the standard deviation of the original population in each case (σ for numerical data, and $p[1-p]$ for categorical data). Therefore, as the population standard deviation (the numerator) increases, the standard error (the entire fraction) also increases. More variability in the population means more variability in the sample means (or in the sample proportions). This increased variability can be offset by increasing the sample size, because as n (the denominator) increases, the overall fraction comprising the standard error decreases.

Anyone can plug numbers into a formula and report a measure of what they feel (or want you to believe) is the true accuracy of their results. But if those results are biased to begin with, their accuracy isn't relevant. (The formulas don't know this, though, so you need to be on the lookout.) Be sure to check to see how the sample in a particular study was selected and how the data were collected before examining any measures of how much those results are expected to vary. (Chapter 17 covers these issues in greater detail.)

Chapter 10

Leaving Room for a Margin of Error

Good survey and experiment researchers always include some measure of how accurate their results are, so that consumers of the information they generate can put the results into perspective. This measure is called the *margin of error* — it's a measure of how close the sample result should be to the population parameter being studied.

Thankfully, many journalists are also realizing the importance of the margin of error in assessing information, so reports that include the margin of error are beginning to appear in the media. But what does the margin of error really mean, and does it tell the whole story?

This chapter looks at the margin of error and what it can and can't do to help you assess the accuracy of statistical information. It also examines the issue of sample size; you may be surprised at how small a sample can be used to get a good handle on the pulse of America — or the world — if the research is done correctly.

Exploring the Importance of That Plus or Minus

The margin of error is a term that you may have heard of before, most likely in the context of survey results. For example, you may have heard someone report, "This survey had a margin of error of plus or minus three percentage points." You may have wondered what you're supposed to do with that information, and how important it really is. The truth is, the survey results themselves (with no margin of error) are only a measure of how the *sample* of selected individuals felt about the issue; they don't reflect how the *entire population* may have felt, had they all been asked (what a job that would be!). The margin of error helps you measure how close you are to the truth about the entire population being studied, while still using only the information gathered from a sample of that population.

As discussed in Chapter 3, a sample is a representative group taken from the population you're interested in studying. Results based on a sample won't be exactly the same as what you would've found for the entire population, because when you take a sample, you don't get information from everyone in the population. But if the study is done right (see Chapter 17 for more about designing good studies), the results from the sample should be close to the actual values for the entire population.

The margin of error doesn't mean someone made a mistake; all it means is that you didn't get to sample everybody in the population, so you expect your sample results to be "off" by a certain amount. In other words, you acknowledge that your results could change with subsequent samples, and are only accurate to within a certain range, which is the margin of error.

Consider one example of the type of survey conducted by some of the leading polling organizations, such as The Gallup Organization. Suppose its latest poll sampled 1,000 people from the United States, and the results show that 520 people (52%) think the president is doing a good job, compared to 48% who don't think so. Suppose Gallup reports that this survey had a margin of error of plus or minus 3%. Now, you know that the majority of the people in this *sample* approve of the president, but can you say that the majority of *all Americans* approve of the president? In this case, you can't. Why not?

If 52% of *those sampled* approve of the president, you can expect that the percent of the *population of all Americans* who approve of the president will be 52%, plus or minus 3%. So you can say that somewhere between 49% and 55% of all Americans approve of the president. That's as close as you can get with your sample of 1,000. But notice that 49%, the lower end of this range, represents a minority, because it's less than 50%. So you really can't say that a majority of the American people support the president, based on this sample. You can only say that between 49% and 55% of all Americans support the president, which may or may not be a majority.

Think about the sample size for a moment. Isn't it interesting that a sample of only 1,000 Americans out of a population of more than 288,000,000 can lead you to be within plus or minus only 3% on your survey results? That's incredible! So for large populations, to get a really good idea of what's happening, you need to sample only a very tiny amount of the total. Statistics is indeed a powerful tool for finding out how people feel about issues. That's probably why so many people conduct surveys, and why you're so often bothered about responding to them.

For a quick-and-dirty way to get a rough idea of what the margin of error is for any given sample size, simply take the sample size *(n)*, and then find 1 divided by the square root of *n*. For the Gallup poll example, *n* = 1,000 and its square root is roughly 31.62, so the margin of error is roughly 1 divided by 31.62, or about 0.03, which is equivalent to 3%. In the remainder of this chapter, you see how to get a more accurate measure of the margin of error.

Finding the Margin of Error: A General Formula

The margin of error is the amount of "plus or minus" that is attached to your sample result when you move from discussing the sample itself to discussing the whole population that it represents. Therefore, you know that the general formula for the margin of error contains a "±" in front of it. So, how do you come up with that plus or minus amount (other than taking a rough estimate, as shown above)? This section shows you how.

Measuring sample variability

Sample results vary, but by how much? According to the central limit theorem (see Chapter 9), when sample sizes are large enough, the distribution of the sample proportions (or the sample averages) will follow a bell-shaped curve (or normal distribution — see Chapter 8). Some of the sample proportions (or sample averages) will overestimate the population value and some will underestimate it, but most will be close to the middle. And what's the middle? If you averaged out the results from all of the possible samples you could take, the average would be the real population proportion, in the case of categorical data, or real the population average, in the case of numerical data. Normally, you don't have the time or the money to look at all of the possible sample results and average them out, but knowing something about all of the other sample possibilities does help you to measure the amount by which you expect your one sample proportion (or average) to vary.

Standard errors are the basic building block of the margin of error. The *standard error* of a statistic is basically equal to the standard deviation of the data divided by the square root of *n* (the sample size). This reflects the fact that the sample size greatly affects how much that sample statistic is going to vary from sample to sample. (See Chapter 9 for more about standard errors.)

The number of standard errors you wish to add or subtract to get the margin of error depends on how confident you wish to be in your results (this is called your *confidence level*). Typically, you want to be about 95% confident, so the basic rule is to add or subtract about 2 standard errors (1.96, to be exact) to get the margin of error. This allows you to account for about 95% of all possible results that may have occurred with repeated sampling. To be 99% confident, you add and subtract about 3 standard errors (2.58, to be exact). (See Chapter 12 for more discussion on confidence levels and number of standard errors.)

To be *exact* about the number of standard errors you wish to add or subtract in order to calculate the margin of error for any confidence level, you need to use a special bell-shaped curve called the *standard normal distribution*. (See Chapter 8 for details.) For any given confidence level, a corresponding value on the standard normal distribution (called a *Z-value*) represents the number of standard errors to add and subtract to account for that confidence level. For 95% confidence, the exact Z-value is 1.96 (which is "about" 2), and for 99% confidence, the exact Z-value is 2.58 (which is "about" 3). Some of the more commonly used confidence levels, along with their corresponding Z-values, are given in Table 10-1.

Table 10-1	Z-Values for Selected (Percentage) Confidence Levels
Percentage Confidence	*Z-Value*
80	1.28
90	1.64
95	1.96
98	2.33
99	2.58

Calculating margin of error for a sample proportion

When the polling question asks people to choose from a range of answers (for example, "Do you approve, disapprove, or have no opinion about the president's performance?"), the statistic used to report the results is the proportion of people from the sample who fell into each group, otherwise known as the *sample proportion*, or *sample percentage*. The general formula for margin of error for your sample proportion is $Z \times \sqrt{\dfrac{\hat{p}(1-\hat{p})}{n}}$, where $\hat{p}$ is the sample proportion, n is the sample size, and Z is the appropriate Z-value for your desired level of confidence (from Table 10-1).

Here are the steps for calculating the margin of error for a sample percentage:

1. **Find the sample proportion, $\hat{p}$, and the sample size, n.**

2. **Multiply $\hat{p} \times (1 - \hat{p})$.**

3. **Divide the result by n.**

4. **Take the square root of the calculated value.**

   You now have the standard error.

5. **Multiply the result by the appropriate Z-value for the confidence level desired.**

   Refer to Table 10-1. The Z-value is 1.96 if you want to be about 95% confident in your results.

Looking at an example involving whether Americans approve of the president, you can find the margin of error. First, assume you want a 95% level of confidence, so Z = 1.96. The number of Americans in the sample who said they approve of the president was found to be 520. This means that the sample proportion, $\hat{p}$ is 520 ÷ 1,000 = 0.52. (The sample size, n, was 1,000.) The margin of error for this polling question is calculated in the following way:

$$Z \times \sqrt{\frac{\hat{p}(1-\hat{p})}{n}} = 1.96 \times \sqrt{\frac{(0.520)(1-0.520)}{1,000}} = 1.96 \times \sqrt{\frac{0.2496}{1,000}}$$

$$= 1.96 \times \sqrt{0.0002} = 1.96 \times 0.0158 = 0.0310 = 3.10\%.$$

A sample proportion is the decimal version of the sample percentage. In other words, if you have a sample percentage of 5%, you must use 0.05 in the formula, not 5. To change a percentage into decimal form, simply divide by 100. After all your calculations are finished, you can change back to a percentage by multiplying your final answer by 100%.

Reporting results

To report the results from this poll, you would say, "Based on my sample, 52% of all Americans approve of the president, plus or minus a margin of error of 3.1%." (Hey, you sound almost as good as Gallup!)

How does a polling organization report its results? Here is basically how it's done:

Based on the total sample of adults in (this) survey, we are 95% confident that the margin of error for our sampling procedure and its results is no more than ±3.1 percentage points.

It sounds in a way like that long list of disclaimers that comes at the end of a car-leasing advertisement. But now you can understand the fine print!

Never accept the results of a survey or study without the margin of error for the study. The margin of error is the only way to measure how close the sample information is to the actual population you're interested in. Sample results vary, and if a different sample had been chosen, a different sample result may have been obtained; you need the margin of error to be able to say how close the sample results are to the actual population values. The next time you hear a media story about a survey or poll that was conducted, take a closer look to see if the margin of error is given. Some news outlets are getting better about reporting the margin of error for surveys, but what about other studies?

Calculating margin of error for a sample average

When a research question asks people to give a numerical value (for example, "How many people live in your house?"), the statistic used to report the results is the average of all the responses provided by people in the sample, otherwise known as the *sample average*.

The general formula for margin of error for your sample average is $Z \times \dfrac{s}{\sqrt{n}}$, where s is the sample standard deviation, *n* is the sample size, and Z is the appropriate Z-value for your desired level of confidence (from Table 10-1).

Here are the steps for calculating the margin of error for a sample average:

1. **Find the sample standard deviation, *s*, and the sample size, *n*.**

   For more information on how to calculate the average and standard deviation, see Chapter 5.

2. **Divide the sample standard deviation by the square root of the sample size.**

   You now have the standard error.

3. Multiply by the appropriate Z-value (refer to Table 10-1).

The Z-value is 1.96 if you want to be about 95% confident.

For example, suppose you're the manager of an ice cream shop, and you're training new employees to be able to fill the large-size cones with the proper amount of ice cream (10 ounces each). You want to estimate the average weight of the cones they make, including a margin of error. You ask each of your new employees to randomly spot check the weights of the large cones they make and record those weights on a notepad. For $n = 50$ cones sampled, the average was found to be 10.3 ounces, with a sample standard deviation of $s = 0.6$ ounces. What's the margin of error? (Assume you want a 95% level of confidence.) It would be calculated this way:

$$Z \times \frac{s}{\sqrt{n}} = 1.96 \times \frac{0.6}{\sqrt{50}} = 0.17$$

So, to report these results, you would say that based on the sample of 50 cones, you estimate that the average weight of all large cones made by the new employees to be 10.3 ounces, with a margin of error of plus or minus 0.17 ounces.

Notice in the ice-cream-cones example, the units are ounces, not percentages! When working with and reporting results about data, always remember what the units are. Also be sure that statistics are reported with their correct units of measure, and if they're not, ask what the units are.

To avoid round-off error in your calculations, keep at least two non-zero digits after the decimal point throughout each step of the calculations. Round-off errors tend to accumulate, and you can be way off by the time you're finished if you round off too soon.

Being sure you're right

If you want to be more than 95% confident about your results, you need to add and subtract more than 2 standard errors. For example, to be 99% confident, you would add and subtract about 3 standard errors to obtain your margin of error. This makes the margin of error larger, which is generally not a good thing. Most people don't think adding and subtracting another whole standard error is worthwhile, just to be 4% more confident (99% versus 95%) in the results obtained. The only way to be 100% sure of your results is to make your margin of error so large (by adding or subtracting many, many standard errors) that it covers every single possibility. For example, you may end up saying something like "I'm 100% sure that the percentage

of people in the population who like ice cream is 50%, plus or minus 50%."
In such a case, you would be 100% sure of your results, but what would they
mean? Nothing.

You can never be completely certain that your sample results do reflect the
population, even with the margin of error included (unless you do something
crazy like include 100% of all the possibilities as in the preceding ice-cream
example). Even if you're 95% confident in your results, that actually means
that if you repeat the sampling process over and over, 5% of the time, the
sample won't represent the population well, simply due to chance (not
because of problems with the sampling process or anything else). So, all
results need to be viewed with that in mind. After all, statistics means never
having to say you're certain!

Determining the Impact of Sample Size

The two most important ideas regarding sample size are the following:

✔ All these formulas work well as long as the sample size is large enough.
(So, how large is "large enough"? This is covered in this section.)

✔ Sample size and margin of error have an inverse relationship.

This section illuminates both concepts.

How large is large enough?

Almost all surveys are done on hundreds or thousands of people, and that is
generally a large enough sample of people to make the theory behind the statistical formulas all work out. However, statisticians have worked out several
general rules to be sure that the sample sizes are large enough.

For sample proportions, you need to be sure that $n \times \hat{p}$ is at least 5, and
$n \times (1 - \hat{p})$ is at least 5. For example, in a preceding example of a poll on the
president, $n = 1,000$, $\hat{p} = 0.52$ and $1 - 0.52 = 0.48$. So, $n \times \hat{p} = 1,000 \times 0.52 = 520$,
and $n \times (1 - \hat{p}) = 1,000 \times 0.48 = 480$. Both of these are safely above 5, so everything is okay.

For sample averages, you need only look at the sample size itself. In general,
the sample size, n, should be above about 30 for the statistical theory to
hold. Now, if it's 29, don't panic; 30 is not a magic number, it's just a general
rule.

Sample size and margin of error

The relationship between margin of error and sample size is simple: As the sample size increases, the margin of error decreases. This is an inverse relationship because the two move in opposite directions. If you think about it, it makes sense that the more information you have, the more accurate your results are going to get. (That, of course, assumes that the data were collected and handled properly.)

If you're interested in the math, I explain more about this inverse relationship in Chapter 9.

Bigger isn't always (that much) better!

In the preceding example of the poll involving the approval rating of the president, the results of a sample of only 1,000 people from all 288,000,000 residents in the United States could get to within 3% of what the whole population would have said, if they had all been asked. How does that work?

Using the formula for margin of error for a sample proportion, you can look at how the margin of error changes dramatically for samples of different sizes.

Suppose in the presidential approval example that $n = 500$. (Recall that $\hat{p} = 0.52$ for this example.) Therefore the margin of error for 95% confidence is

$$Z \times \sqrt{\frac{\hat{p}(1 - \hat{p})}{n}} = 1.96 \times \sqrt{\frac{(0.520)(0.480)}{500}} = 1.96 \times 0.0223 = 0.0438,$$ which is

equivalent to 4.38%. When $n = 1,000$ in the same example, the margin of error

(for 95% confidence) is $1.96 \times \sqrt{\frac{(0.520)(0.480)}{1,000}} = 1.96 \times 0.0158 = 0.0310,$

which is equal to 3.10%. If n were increased to 1,500, the margin of error

(with the same level of confidence) becomes $1.96 \times \sqrt{\frac{(0.520)(0.480)}{1,500}} =$

$1.96 \times 0.0129 = 0.0253,$ or 2.53%. Finally, when $n = 2,000$, the margin of error

is $1.96 \times \sqrt{\frac{(0.520)(0.480)}{2,000}} = 1.96 \times 0.0112 = 0.0219,$ or 2.19%.

Looking at these different results, you can see that after a certain point, you have a diminished return for larger and larger sample sizes. Each time you survey one more person, the cost of your survey increases, and going from a sample size of 1,500 to a sample size of 2,000 decreases your margin of error by only 0.34% (one third of one percent!). The extra cost and trouble to get that small decrease in the margin of error may not be worthwhile. Bigger isn't always that much better!

But what may really surprise you is that bigger isn't always even a little bit better; bigger can actually be worse! I explain this in the following section.

Limiting the Margin of Error

The margin of error is a measure of how close you expect your sample results to represent the entire population being studied. (Or at least it gives an upper limit for the amount of error you should have.) Because you're basing your conclusions about the population on your one sample, you have to account for how much those sample results could vary, just due to chance.

Another view of margin of error is that it represents the maximum *expected* distance between the sample results and the actual population results (if you'd been able to obtain them through a census). Of course if you had the absolute truth about the population, you wouldn't be trying to do a survey, would you?

Just as important as knowing what the margin of error measures is realizing what the margin of error does *not* measure. The margin of error does *not* measure anything other than chance variation. That is, it doesn't measure any bias or errors that happen during the selection of the participants, the preparation or conduct of the survey, the data collection and entry process, or the analysis of the data and the drawing of the final conclusions.

Bigger samples don't always mean better samples! A good slogan to remember when examining statistical results is "garbage in equals garbage out." No matter how nice and scientific the margin of error may look, remember that the formula that was used to calculate it doesn't have any idea of the quality of the data that the margin of error is based on. If the sample proportion or sample average was based on a *biased sample* (one that favored certain people over others), a bad design, bad data-collection procedures, biased questions, or systematic errors in recording, then calculating the margin of error is pointless, because it won't mean a thing. For example, a total 50,000 people surveyed sounds great, but if they were all visitors to a certain Web site, the margin of error for this result means nothing, because the calculation is all based on biased, bogus results! Of course, some people go ahead and report it anyway, so you have to find out what went into the formula: good information or garbage? If it turns out to be garbage, you know what to do about the margin of error. Ignore it. (For more information on errors that can take place during a survey or experiment, see Chapters 16 and 17, respectively.)

The Gallup Organization addresses the issue of what margin of error does and doesn't measure in a disclaimer that it uses to report its survey results. The organization tells you that besides sampling error, surveys can have additional errors or bias due to question wording and some of the logistical issues involved in conducting surveys (such as missing data due to phone numbers that are no longer current). This means that even with the best of intentions and the most meticulous attention to details and process control, stuff happens. Nothing is ever perfect. But what you need to know is that the margin of error can't measure the extent of those other types of errors.

Part V
Guesstimating with Confidence

The 5th Wave   By Rich Tennant

STATISTICIAN CONVENTION

"Be careful. He looks like a good husband, but there's a histogram going around of his marriages and divorces that doesn't look very good."

In this part . . .

Anytime someone gives you a statistic by itself, he or she hasn't really given you the full story. The statistic alone is missing the most important part: by how much that statistic is expected to vary. All good estimates contain not just a statistic but also a margin of error. This combination of a statistic plus or minus a margin of error is called a confidence interval. Confidence intervals go beyond a single statistic; instead, they offer important information about the accuracy of the estimate.

This part gives you a general, intuitive look at confidence intervals: their function, formulas, calculations, influential factors, and interpretation. You also get quick references and examples for the most commonly used confidence intervals.

Chapter 11

The Business of Estimation: Interpreting and Evaluating Confidence Intervals

. . . .

In This Chapter

▶ Seeing how the estimation process works (or at least how it *should* work)

▶ Getting a general formula for a confidence interval

▶ Interpreting the results of a confidence interval

▶ Detecting misleading results

. . . .

Most statistics are used to estimate some characteristic about a population of interest, such as average household income, the percentage of people who buy Christmas gifts online, or the average amount of ice cream consumed in the United States every year (maybe that statistic is better left unknown). Such characteristics of a population are called *parameters*. Typically, people want to estimate (take a good guess at) the value of a parameter by taking a sample from the population and using statistics from the sample that will give them a good estimate. The question is, how do you define "good estimate"?

The best guess would be no guess at all — go out and actually come up with the parameter itself. You can't determine the exact value of a parameter without conducting a census of the entire population — a daunting and expensive task in most cases. Statisticians, however, remain unfazed by the challenge and often say, "Being a statistician means never having to say you're certain; you only have to get close." Of course, statisticians want to be confident that their results are as close as they can be to the truth, within a certain time frame and budget, and "close" is easier to accomplish than you may think. As long as the process is done correctly (and in the media, it often isn't!), an estimate can get very close to the parameter. This chapter gives you an overview of confidence intervals (the type of estimates used and recommended by statisticians), why they should be used (as opposed to just a one-number estimate), how to interpret a confidence interval, and how to spot misleading estimates.

Realizing That Not All Estimates Are Created Equal

Read any magazine or newspaper or listen to any newscast, and you hear a number of statistics, many of which are estimates of some quantity or another. You may wonder how they came up with those statistics: In some cases, the numbers are well researched; in other cases, they're just a shot in the dark. Here are some examples of estimates that I came across in one single issue of a leading business magazine. They come from a variety of sources:

- 26 million folks play golf at least once a year.

- 6.7 percent of U.S. homes were purchased without a down payment.

- Even though some jobs are harder to get these days, some areas are really looking for recruits: Over the next eight years, 13,000 nurse anesthetists will be needed. Pay starts from $80,000 to $95,000.

- The average number of bats used by a Major League baseball player per season is 90.

- 7.4 million U.S. residents took cruise vacations in 2002. Of those people, about 4 percent of them visited the ship's medical staff.

- The Lamborghini Murcielago can go from 0 to 60 mph in 3.7 seconds with a top speed of near 205 mph.

Some of these estimates are easier to obtain than others. Here are some observations I was able to make about those estimates:

- How do you know that 26 million people play golf once at least once a year? Actually, this one may not be too hard to get, because golfers must always sign in whenever they play at a golf course. So, a survey of golf course sign-in sheets could give a good estimate of how many people play at least once a year. (The only hard part would be not double-counting people you've counted before on previous sign-in sheets.)

- A survey may be able to estimate the percentage of cruisers who need medical attention or the percentage of homes purchased without a down payment. If the survey is done correctly (see Chapter 16), these data are probably pretty accurate.

✔ How do you estimate how many nurse anesthetists are needed over the next eight years? You can start by looking at how many will be retiring in that time; but that won't account for growth. A prediction of the need in the next year or two would be close, but eight years into the future is much harder to do.

✔ The average number of bats used per Major League baseball player in a season could be found by surveying the players themselves, the people who take care of their equipment, or the bat companies that supply the bats.

✔ Determining car speed is more difficult, but could be conducted as a test with a stopwatch. And they should use many different cars (not just one) of the same make and model.

Not all statistics are created equal. To determine whether a statistic is reliable and credible, don't just take it at face value. Think about whether it makes sense and how you would go about formulating an estimate. If the statistic is really important to you, find out what process was used to come up with the estimate.

Linking a Statistic to a Parameter

The U.S. Census Bureau estimates the *median* household income for the United States and breaks it down by each state in its yearly report from the Current Population Survey. Why estimate the median (middle number) and not the mean (average) household income? (See Chapter 5 for more discussion on the mean versus the median.) Because household income tends to be skewed, with many folks at the lower end of the income spectrum, and fewer individuals at the extreme upper end.

To estimate the median household income, the Census Bureau takes a random sample of about 28,000 households and asks questions (household income is among those questions). Based on the sample data of 28,000 homes, the Bureau calculates the median household income for this sample: For the year 2000, the sample median household income was $42,228.

The Census Bureau uses the sample median household income (a statistic) to estimate the median household income for the whole U.S. (the parameter). Yet because the Bureau knows that a sample can't possibly reflect the entire population completely accurately, it includes a *margin of error* (see Chapter 10) with the results. This plus or minus (±) that's added to any estimate, helps put the results into perspective. When you know the margin of error, you have an idea of how much error may be in the estimate, due to the fact that it was based on a sample of the population and not on the entire population.

Because the Bureau didn't sample the entire population and knows that the sample may not represent the entire population perfectly, the Bureau calculates a margin of error for the median of the sample group and includes that as part of the estimate. For the year 2000, the margin of error of the sample median household income was $258. The U.S. Census Bureau therefore estimates that in 2000, the median household income was $42,228, plus or minus $258 or $42,228 ± $258. This represents the confidence interval for the median household income of the United States (see the following section about confidence intervals).

Note the margin of error is fairly small in the above example; that's because of the high sample size used (you get what you pay for!). See Chapter 10 for more information on the relationship between sample size and the size of the margin of error.

Making Your Best Guesstimate

The best way to estimate a parameter (a characteristic of an entire population) is to come up with a statistic, plus or minus a margin of error, that's based on a large sample. In this way, your result presents an estimate based on your sample, along with some indicator of how much that estimate could vary from sample to sample.

A statistic plus or minus a margin of error is called a *confidence interval:*

- ✔ The word *interval* is used because your result becomes an interval. For example, say the percentage of kids who like baseball is 40 percent, plus or minus 3.5 percent. That means the percentage of kids who like baseball is somewhere between 40% – 3.5% = 36.5% and 40% + 3.5% = 43.5%. Thus, the lower end of the interval is your statistic minus the margin of error, and the upper end is your statistic plus the margin of error.

- ✔ The word *confidence* is used because you have a certain amount of confidence in the process by which you got your interval. This is called your *confidence level.*

You can find formulas and examples for the most commonly used confidence intervals in Chapter 13.

Interpreting Results with Confidence

Suppose you, a research biologist, are trying to catch a fish using a hand net, and the size of your net represents the width of your confidence interval. (The width is twice the margin of error, due to adding and subtracting.)

Suppose your confidence level is 95%. What does this really mean? It means that if you scoop this particular net into the water over and over again, you'll catch a fish 95% of the time. Catching a fish here means your confidence interval was correct and contains the true parameter (in this case the parameter is represented by the fish itself).

But does this mean that on any given try you have a 95% chance of catching a fish? No. Is this confusing? It certainly is. Here's the scoop (no pun intended). On a single try, say you close your eyes before you scoop your net into the water. At this point, your chances of catching a fish are 95%. But then go ahead and scoop your net through the water with your eyes still closed. After that's done, however, you have only two possible outcomes; you either caught a fish or you didn't; probability isn't involved.

Likewise, after data have been collected and the confidence interval has been calculated, you either captured the true population parameter or you didn't. So you're not saying you're 95% confident that the parameter is in the interval, because you either captured it or you didn't. What you are 95% confident about is the process by which you're getting the data and forming the confidence interval in the long run. You know that this process will result in intervals that capture the mean 95% of the time. The other 5% of the time, the data collected in the sample just by chance had abnormally high or low values in it and didn't represent the population. In those cases, you won't capture the parameter.

So you know that with the size and composition of your net, you're going to catch a fish 95% of the time. On any given try, however, you either catch a fish or you don't.

 Confidence level, sample size, and population variability all play a role in influencing the size of the margin of error and the width of a confidence interval. But the margin of error, and hence the width of a confidence interval, is meaningless if the data that went into the study were biased and/or unreliable. The best advice is to look at how the data were collected before accepting a reported margin of error as the truth (see Chapter 10).

Spotting Misleading Confidence Intervals

When data come from well-designed surveys or experiments (see Chapters 16 and 17) and are based on large random samples (see Chapter 9), you can feel good about the quality of the information. When the margin of error is small, relatively speaking, you would like to say that these confidence intervals provide accurate and credible estimates of their parameters. This is not always the case, however.

Not all estimates are as accurate and reliable as some may want you to think. For example, a Web site survey result based on 20,000 hits may have a small margin of error according to the formula, but the margin of error means nothing if the survey is only given to people who happened to visit that Web site. In other words, the sample isn't even close to being a random sample (where everyone in the population has an equal chance of being chosen to participate). Nevertheless, such results do get reported, along with their margins of error that make the study seem truly scientific. Beware of these bogus results! (See Chapter 10 for more on the limits of the margin of error.)

Before making any decisions based on someone's estimate, do the following:

✔ Investigate how the statistic was created; it should be the result of a scientific process that results in reliable, unbiased, accurate data. (See Chapters 2 and 3.)

✔ Look for a margin of error. If one isn't reported, go to the original source and request it.

✔ Remember that if the statistic isn't reliable or contains bias, the margin of error will be meaningless. (See Chapter 16 for avoiding bias in survey data and see Chapter 17 for criteria for good data in experiments.)

Chapter 12

Calculating Accurate Confidence Intervals

A *confidence interval* is a fancy phrase for a statistic that has a margin of error to go along with it (see Chapter 11 for a basic overview of confidence intervals; see Chapter 10 for information on margin of error). Because most statistics are calculated for the purpose of estimating characteristics of a population (called *parameters*) from a sample, every statistic should include a margin of error as a measure of its level of accuracy. After all (as Chapter 9 says), when you're taking samples, results vary!

In this chapter, you find out how to calculate your own confidence interval. You also get the lowdown on some of the finer points of confidence intervals: what makes them narrow or wide, what makes you more or less confident in their results, and what they do and don't measure. With this information, you know what to look for when being presented statistical results, and you understand how to gauge the true accuracy of those results.

Calculating a Confidence Interval

A confidence interval is composed of a statistic, plus or minus a margin of error (see Chapter 10). For example, suppose you want to know the percentage of vehicles in the United States that are pickup trucks (that's the parameter, in this case). You can't look at every single vehicle in the United States, so you take a random sample of 1,000 vehicles over a range of highways at different times of the day. You find that 7% of the vehicles in your sample are pickup trucks. Now, you don't want to say that *exactly* 7% of all vehicles on U.S. roads are pickup trucks, because you know this is only based on the 1,000 vehicles you sampled. While you hope that 7% is close to the true percentage, you can't be sure because you based your results on a sample of vehicles, not on all of the vehicles in the United States.

So what to do? You add and subtract a margin of error to indicate how much error you expect your sample result to have. (See Chapter 10 for more on the margin of error.) The error isn't due to anything you did wrong, it simply comes from the fact that a *sample* (a study of a portion of the population), not a *census* (a study of the entire population), was done.

The *width* of your confidence interval is two times the margin of error. For example, suppose the margin of error was 5%. A confidence interval of 7%, plus or minus 5%, goes from 7% – 5% = 2%, all the way up to 7% + 5% = 12%. That means it has a width of 12% – 2% = 10%. A simpler way to calculate this is to say that the width of the confidence interval is two times the margin of error. In this case, the width of the confidence interval is $2 \times 5\% = 10\%$.

The width of a confidence interval is the distance from the lower end of the interval (statistic – margin of error) to the upper end of the interval (statistic + margin of error). You can always calculate the width of a confidence interval quickly by taking 2 times the margin of error.

The following are the general steps for estimating a parameter with a confidence interval, along with references where you can find more detailed information on how to accomplish each step.

1. **Choose your confidence level and your sample size (see Chapter 9).**

2. **Select a random sample of individuals from the population (see Chapter 3).**

3. **Collect reliable and relevant data from the individuals in the sample.**

   See Chapter 16 for survey data and Chapter 17 for data from experiments.

4. **Summarize the data into a statistic, usually a mean or proportion (see Chapter 5).**

5. **Calculate the margin of error (see Chapter 10).**

6. **Take the statistic plus or minus the margin of error to get your final estimate of the parameter.**

   This is called a *confidence interval* for that parameter.

Teen attitudes toward smokeless tobacco

An ongoing study conducted by the University of Michigan monitors teenagers' attitudes on a number of issues, including the perceived risk of smokeless tobacco (commonly called chewing tobacco). The study shows that more of today's teenagers perceive smokeless tobacco to cause great risk compared to 15 years ago. Their results are the following.

✔ In a sample in 2001 of 2,100 twelfth graders, 45.4% perceived smokeless tobacco to cause a great risk for harm. The margin of error was plus or minus 2%.

A 95% confidence interval for the percentage of *all* twelfth graders who perceive smokeless tobacco to be of great risk is therefore 45.4% ± 2%.

✔ Based on a sample of 3,000 twelfth graders in 1986, the confidence interval for all twelfth graders who felt that smokeless tobacco caused a great risk was 25.8% ± 1.6%.

Choosing a Confidence Level

Notice that the example showing teen attitudes towards tobacco (see the "Teens attitude toward smokeless tobacco" sidebar) includes the phrase a "95% confidence interval." Every confidence interval (and every margin of error, for that matter) has a confidence level associated with it. In that example, the confidence level was 95%. A confidence level helps you account for the other possible sample results you could have gotten, when you're making an estimate of a parameter using the data from only one sample. If you want to account for 95% of the other possible results, your confidence level would be 95%.

Variability in sample results is measured in terms of number of standard errors. A *standard error* is similar to the standard deviation of a data set, only a standard error applies to sample means or sample percentages that you could have gotten, if different samples were taken. (See Chapter 10 for information on standard errors.) Every confidence level has a corresponding number of standard errors that have to be added or subtracted. This number of standard errors is called the Z-value (because it corresponds to the standard normal distribution). See Table 10-1 in Chapter 10.

What level of confidence level is typically used by researchers? I've seen confidence levels ranging from 80% to 99%. The most common confidence level is 95%. In fact, statisticians have a saying that goes, "Why do statisticians like their jobs? Because they have to be correct only 95% of the time." (Tacky, but sort of catchy, isn't it?)

Being 95% confident means that if you take many, many samples, and calculate a confidence interval each time, based on the results, 95% of those samples will result in confidence intervals that are right on target and actually contain the true parameter. In order to have a 95% confidence level, the empirical rule says that you need to add and subtract "about" 2 standard errors. The central limit theorem allows you to be more exact about the amount, so the "about 2" actually becomes 1.96. See Table 10-1 in Chapter 10 for selected confidence levels and their corresponding Z-values.

If you want to be more than 95% confident about your results, you need to add and subtract more than two standard errors. For example, to be 99% confident, you would add and subtract about three standard errors to obtain your margin of error. The higher the confidence level, the larger the Z-value, the larger the margin of error, and the wider the confidence interval (assuming everything else stays the same). You have to pay a certain price for more confidence.

Note I said "assuming everything else stays the same." You can offset an increase in the margin of error by increasing the sample size. See the "Factoring in the Sample Size" section for more on this.

Zooming In on Width

The ultimate goal when making an estimate using a confidence interval is to have the confidence interval be narrow. That means you're zooming in on what the parameter is. Having to add and subtract a large amount only makes your result much less accurate. For example, suppose you're trying to estimate the percentage of semi trucks on the interstate between the hours of 12 a.m. and 6 a.m., and you come up with a 95% confidence interval that claims the percentage of semis is 50%, plus or minus 50%. Wow, that narrows it down! (Not.) You've defeated the purpose of trying to come up with a good estimate.

In this case, the confidence interval is much too wide. You'd rather say something like this: A 95% confidence interval for the percentage of semis on the interstate between 12 a.m. and 6 a.m. is 50%, plus or minus 3%. This would require a much larger sample size, but that would be worthwhile.

So, if a small margin of error is good, is smaller even better? Not always. To get an extremely narrow confidence interval, you have to conduct a much more difficult — and expensive — study, so a point comes where the increase in price doesn't justify the marginal difference in accuracy. Most people are pretty comfortable with a margin of error of 2% to 3% when the estimate itself is a percentage (like the percentage of women, Republicans, or smokers).

A narrow confidence interval is a good thing.

How do you go about ensuring that your confidence interval will be narrow enough? You certainly want to think about this issue before collecting your data; after the data are collected, the width of the confidence interval is set.

Three factors affect the width of a confidence interval:

- ✔ The confidence level (as discussed in the preceding section)
- ✔ The sample size
- ✔ The amount of variability in the population

The formula for the margin of error associated with a sample mean is $Z \times \frac{s}{\sqrt{n}}$, where:

- ✔ Z is the value from the standard normal distribution corresponding to the confidence level (see Table 10-1 in Chapter 10).
- ✔ n is the sample size (see Chapter 9).
- ✔ $\frac{s}{\sqrt{n}}$ is the standard error of the sample mean. (See Chapter 10 for more on standard error.)

A confidence interval for the average would then be $\bar{x}$ plus or minus the margin of error. Chapter 13 gives the formulas for the most common confidence intervals you're likely to come across.

Each of these three factors (confidence level, sample size, and population variability) plays an important role in influencing the width of a confidence interval. You've already seen the effects of confidence level. In the following section, you explore how sample size and population variability affect the width of a confidence interval.

Note that the sample statistic itself (for example, 7% of vehicles in the sample are pickup trucks) isn't related to the width of the confidence interval. Instead, the margin of error and the three factors involved in it are totally responsible for determining the width of a confidence interval.

Factoring In the Sample Size

The relationship between margin of error and sample size is simple: As the sample size increases, the margin of error decreases. This confirms what you hope is true: The more information you have, the more accurate your results are going to be. (That of course, assumes that the information is good, credible information. See Chapter 2 for how statistics can go wrong.)

Looking at the formula for margin of error for the sample mean, notice that it has an n in the denominator of a fraction (this is the case for most any margin of error formula): $Z \times \frac{s}{\sqrt{n}}$. As n increases, the denominator of this fraction increases, which makes the overall fraction get smaller. That makes the margin of error smaller and results in a narrower confidence interval.

When you need a high level of confidence, you have to increase the Z-value and, hence, margin of error, resulting in a wider confidence interval, which isn't good. But, you can offset this wider confidence interval by increasing the sample size and bringing the margin of error back down, thus narrowing the confidence interval. The increase in sample size allows you to still have the confidence level you want, but also ensures that the width of your confidence interval will be small (which is what you ultimately want). You can even determine this information before you start a study: If you know the margin of error you want to get, you can set your sample size accordingly (see Chapter 9).

When your statistic is going to be a percentage (like the percentage of people who prefer to wear sandals during summer), a rough way to figure margin of error is to take 1 divided by the square root of n (the sample size). You can try different values of n and you can see how the margin of error is affected.

Approximately what sample size is needed to have a narrow confidence interval with respect to polls? Using the formula in the preceding paragraph, you can make some quick comparisons. A survey of 100 people will have a margin of error of about $\frac{1}{\sqrt{100}} = 0.10$ or plus or minus 10% (meaning the width of the confidence interval is 20%, which is pretty large.) However, if you survey 1,000 people, your margin of error decreases dramatically, to plus or minus about 3%; the width now becomes only 6%. A survey of 2,500 people results in a margin of error of plus or minus 2% (so the width is down to 4%). That's quite a small sample size to get so accurate, when you think about how large the population is (the U.S. population, for example, is over 280 million!).

Keep in mind, however, you don't want to go too high with your sample size because a point comes where you have a diminished return. For example, moving from a sample size of 2,500 to 5,000 narrows the width of the confidence interval to about $2 \times 1.4 = 2.8\%$, down from 4%. Each time you survey one more person, the cost of your survey increases, so adding another 2,500 people to the survey just to narrow the interval by little more than 1% may not be worthwhile.

Real accuracy depends on the quality of the data as well as on the sample size. A large sample size that has a great deal of bias (see Chapter 2) may appear to have a narrow confidence interval but means nothing. That's like competing in an archery match and shooting your arrows consistently, but finding out that the whole time you're shooting at the next person's target; that's how far off you are. With the field of statistics, though, you can't measure bias, you can only try to minimize it.

The larger the sample size is, the smaller the margin of error will be, and the narrower the confidence interval will get, assuming that everything else stays the same and that the quality of the data is good.

Counting On Population Variability

One of the factors influencing variability in sample results is the fact that the population itself contains variability. If every value in the population were exactly the same, imagine how boring the world would be. (In fact, statisticians wouldn't exist if not for variability.) For example, in a population of houses in a large city like Columbus, Ohio, you see a great deal of variety in not only the types of houses, but also the sizes, and the prices. And the variability in prices of houses in Columbus, Ohio, should be more than the variability in prices of houses in a selected housing development in Columbus.

That means if you take a sample of houses from the entire city of Columbus and find the average price, the margin of error should be larger than if you take a sample from that single housing development in Columbus, even if you have the same confidence level and the same sample size each time. Why? Because the houses in the entire city have more variability in price, and your sample average would change more from sample to sample than it would if you took the sample only from that single housing development, where the prices tend to be very similar. That means you need to sample more houses if you're sampling from the entire city of Columbus in order to have the same amount of accuracy that you would get from that single housing development.

Variability is measured by the standard deviation. The standard deviation of the population (σ) isn't typically known, so you estimate it with s, the standard deviation of the sample (see Chapter 4). Notice that s appears in the numerator of the standard error in the formula for margin of error for the sample mean: $Z \times \dfrac{s}{\sqrt{n}}$. Therefore, as the standard deviation (the numerator) increases, the standard error (the entire fraction) also increases. This results in a larger margin of error and a wider confidence interval.

More variability in the original population increases the margin of error, making the confidence interval wider. This increase can be offset by increasing the sample size.

Chapter 13

Commonly Used Confidence Intervals: Formulas and Examples

*W*henever you want to determine the mean of the population but you can't find it exactly due to time/money constraints (which is usually the case), the next best thing to do is take a sample of the population, find *its* mean, and use that to estimate the mean for the whole population. Then (and see Chapters 11 and 12 for details), you must include some measure of how accurate you expect your sample results to be; after all, you know that those results would change at least a little if you took a different sample. So along with your sample mean, you must include a margin of error (by how much you expect your sample result to change from sample to sample), and your sample mean plus or minus the margin of error combines to form a confidence interval for the population mean.

But figuring the confidence interval can be a little confusing, so in this chapter, I outline the formulas for the four most commonly used confidence intervals (CIs), explain the calculations, and walk you through some examples.

Calculating the Confidence Interval for the Population Mean

When the characteristic that's being measured (such as income, IQ, price, height, quantity, or weight) is *numerical,* most people want to report the mean (average) value for the population, because the average is a one number summary of the population, telling where the center of the population is. You estimate the population mean by using a sample mean, plus or minus a margin of error. The result is called a *confidence interval for the population mean.*

The formula for a CI for a population mean is $\bar{x} \pm Z \times \frac{s}{\sqrt{n}}$, where $\bar{x}$ is the sample mean, s is the sample standard deviation, n is the sample size and Z is the appropriate value from the standard normal distribution for your desired confidence level. (See Chapter 3 for formulas for $\bar{x}$ and s; see Chapter 10 (Table 10-1) for values of Z for given confidence levels.)

To calculate a CI for the population mean (average), do the following:

1. **Determine the confidence level and find the appropriate Z-value.**

   See Chapter 10 (Table 10-1).

2. **Find the sample mean ($\bar{x}$), the sample standard deviation (s), and the sample size (n).**

   See Chapter 3.

3. **Multiply Z times s and divide that by the square root of n.**

   This is the margin of error.

4. **Take $\bar{x}$ plus or minus the margin of error to obtain the CI.**

   The lower end of the CI is $\bar{x}$ minus the margin of error, while the upper end of the CI is $\bar{x}$ plus the margin of error.

For example, suppose you work for the Department of Natural Resources, and you want to estimate, with 95% confidence, the mean (average) length of walleye fingerlings in a fish hatchery pond.

Because you want a 95% confidence interval, your Z is 1.96.

Suppose you take a random sample of 100 fingerlings, and you determine that the average length is 7.5 inches and the standard deviation (s) is 2.3 inches. (See Chapter 4 for calculating the mean and standard deviation.) This means $\bar{x} = 7.5$, $s = 2.3$, and $n = 100$.

Multiply 1.96 times 2.3 divided by the square root of 100 = (10). The margin of error is, therefore, plus or minus $1.96 \times (2.3 \div 10) = 1.96 \times 0.23 = 0.45$ inches.

Your 95% confidence interval for the mean length of walleye fingerlings in this fish hatchery pond is 7.5 inches plus or minus 0.45 inches. (The lower end of the interval is 7.5 − 0.45 = 7.05 inches; the upper end is 7.5 + 0.45 = 7.95 inches.) You can say then, with 95% confidence, that the average length of walleye fingerlings in this entire fish hatchery pond is between 7.05 and 7.95 inches, based on your sample.

When your sample size is small (under 30), a slight modification in your calculations will be needed. This is discussed in Chapter 15.

Determining the Confidence Interval for the Population Proportion

When a characteristic being measured is categorical (for example, opinion on an issue [support, oppose, or are neutral], gender, political party, or type of behavior [do/don't wear a seatbelt while driving]), most people want to report the proportion (or percentage) of people in the population that fall into a certain category of interest. For example, the percentage of people in favor of a four day work week, the percentage of Republicans who voted in the last election, or the proportion of drivers who don't wear seat belts. In each of these cases, the object is to estimate a population proportion using a sample proportion plus or minus a margin of error. The result is called a *confidence interval for the population proportion.*

The formula for a CI for a population proportion is $\hat{p} \pm Z \times \sqrt{\dfrac{\hat{p}(1-\hat{p})}{n}}$, where $\hat{p}$ is the sample proportion, n is the sample size, and Z is the appropriate value from the standard normal distribution for your desired confidence level. (See Chapter 3 for formulas and calculations for $\hat{p}$; see Chapter 10 [Table 10-1] for values of Z for certain confidence levels.)

To calculate a CI for the population proportion:

1. **Determine the confidence level and find the appropriate Z-value.**

   See Chapter 10 (Table 10-1).

2. **Find the sample proportion ($\hat{p}$) by taking the number of people in the sample having the characteristic of interest, divided by the sample size (*n*).**

   Note: $\hat{p}$ should be a decimal value between 0 and 1.

3. **Multiply $\hat{p}$ times (1 – $\hat{p}$), and then divide that amount by *n*.**

4. **Take the square root of the result from Step 3.**

5. **Multiply your answer by Z.**

   This is the margin of error.

6. **Take $\hat{p}$ plus or minus the margin of error to obtain the CI. The lower end of the CI is $\hat{p}$ minus the margin of error and the upper end of the CI is $\hat{p}$ plus the margin of error.**

For example, suppose you want to estimate the percentage of the times you get a red light at a certain intersection.

Because you want a 95% confidence interval, your Z-value is 1.96.

You take a random sample of 100 different trips through this intersection, and you find that you hit a red light 53 times, so $\hat{p} = 53 \div 100 = 0.53$.

Take 0.53 times $(1 - 0.53)$ and divide by 100 to get $0.2491 \div 100 = 0.002491$.

Take the square root to get 0.0499.

The margin of error is, therefore, plus or minus $1.96 \times (0.0499) = 0.0978$.

Your 95% confidence interval for the percentage of times you will ever hit a red light at that particular intersection is 0.53 (or 53%) plus or minus 0.0978 (rounded to 0.10 or 10%). (The lower end of the interval is $0.53 - 0.10 = 0.43$ or 43%; the upper end is $0.53 + 0.10 = 0.63$ or 63%.) In other words, you can say that with 95% confidence, the percentage of the times you should expect to hit a red light at this intersection is somewhere between 43% and 63%, based on your sample.

While performing any calculations involving sample percentages, use the decimal form. After the calculations are finished, convert to percentages by multiplying by 100. To avoid round-off error, keep at least 2 decimal places throughout.

Developing a Confidence Interval for the Difference of Two Means

The goal of many surveys and studies is to compare two populations, such as men versus women, low versus high income families, and Republicans versus Democrats. When the characteristic being compared is numerical (for example, height, weight, or income) the object of interest is the amount of difference in the means (averages) for the two populations. For example, you may want to compare the difference in average age of Republicans versus Democrats, or the difference in average incomes of men versus women. You estimate the difference between two population means by taking a sample from each population and using the difference of the two sample means, plus or minus a margin of error. The result is a *confidence interval for the difference of two population means*.

The formula for a CI for the difference between two population means (averages) is $(\bar{x} - \bar{y}) \pm Z \times \sqrt{\frac{s_1^2}{n_1} + \frac{s_2^2}{n_2}}$, where $\bar{x}$, s_1, and n_1 are the mean, standard deviation and size of the first sample, and $\bar{y}$, s_2, and n_2 are the mean, standard deviation and size of the second sample. Z is the appropriate value from the standard normal distribution for your desired confidence level. (See Chapter 3 for formulas and calculations for means and standard deviations; see Chapter 10 (Table 10-1) for values of Z for certain confidence levels.)

To calculate a CI for the difference between two population means, do the following:

1. **Determine the confidence level and find the appropriate Z-value.**

   See Chapter 10 (Table 10-1).

2. **Find the mean ($\bar{x}$), standard deviation (s_1) and sample size (n_1) of the first sample and the mean ($\bar{y}$), standard deviation (s_2) and sample size (n_2) of the second sample.**

   See Chapter 3.

3. **Find the difference, ($\bar{x} - \bar{y}$), between the sample means.**

4. **Square s_1 and divide it by n_1; square s_2 and divide it by n_2. Add the results together and take the square root.**

5. **Multiply your answer from Step 4 by Z.**

   This is the margin of error.

6. **Take ($\bar{x} - \bar{y}$) plus or minus the margin of error to obtain the CI.**

   The lower end of the CI is ($\bar{x} - \bar{y}$) *minus* the margin of error, while the upper end of the CI is ($\bar{x} - \bar{y}$) *plus* the margin of error.

Suppose you want to estimate with 95% confidence the difference between the mean (average) length of the cobs of two varieties of sweet corn (allowing them to grow the same number of days under the same conditions). Call the two varieties Corn-e-stats and Stats-o-sweet.

Because you want a 95% confidence interval, your Z is 1.96.

Suppose your random sample of 100 cobs of the Corn-e-stats variety averages 8.5 inches, with a standard deviation of 2.3 inches, and your random sample of 110 cobs of Stats-o-sweet averages 7.5 inches, with a standard deviation of 2.8 inches. This means $\bar{x}$ = 8.5, s_1 = 2.3, and n_1 = 100; $\bar{y}$ = 7.5, s_2 = 2.8, and n_2 = 110.

The difference between the sample means, ($\bar{x} - \bar{y}$), from Step 3, is 8.5 – 7.5 = +1 inch. This means the average for Corn-e-stats minus the average for Stats-o-sweet is positive, making Corn-e-stats the larger of the two varieties, in terms of this sample. Is that difference enough to generalize to the entire population, though? That's what this confidence interval is going to help you decide.

Square s_1 (2.3) to get 5.29; divide by 100 to get 0.0529. Square s_2 (2.8) and divide by 110: $7.84 \div 110 = 0.0713$. The sum is $0.0529 + 0.0713 = 0.1242$; the square root of this is 0.3524.

Multiply 1.96 times 0.3524 to get 0.69 inches, the margin of error.

Your 95% confidence interval for the difference between the average lengths for these two varieties of sweet corn is 1 inch, plus or minus 0.69 inches. (The lower end of the interval is $1 - 0.69 = 0.31$ inches; the upper end is $1 + 0.69 = 1.69$ inches.) That means you can say, with 95% confidence, that the Corn-e-stats variety is longer, on average, than the Stats-o-sweet variety, by somewhere between 0.31 and 1.69 inches. (Notice all the values in this interval are positive. That means Corn-e-stats should always on average be longer than Stats-o-sweet, based on your data.)

Notice that you could get a negative value for $(\overline{x} - \overline{y})$. For example, if you had switched the two varieties of corn, you would have gotten –1 for this difference. That's fine; just remember which group is which. A positive difference means the first group has a larger value than the second group; a negative difference means the first group has a smaller value than the second group. If you want to avoid negative values, always make the group with the larger value your first group — all your differences will be positive.

In the case where your sample size is small (under 30), see Chapter 15 for the slight modifications that you need to make to your calculations.

Coming Up with the Confidence Interval for the Difference of Two Proportions

When a characteristic, such as opinion on an issue (support/don't support), of the two groups being compared is *categorical*, people want to report on the differences between the two population proportions — for example, the difference between the proportion of women who support a four-day work week, and the proportion of men who support a four-day work week. You estimate the difference between two population proportions by taking a sample from each population and using the difference of the two sample proportions, plus or minus a margin of error. The result is called a *confidence interval for the difference of two population proportions.*

The formula for a confidence interval for the difference between two population proportions is $\left(\hat{p}_1 - \hat{p}_2 \right) \pm Z \sqrt{\dfrac{\hat{p}_1(1 - \hat{p}_1)}{n_1} + \dfrac{\hat{p}_2(1 - \hat{p}_2)}{n_2}}$, where $\hat{p}_1$ and n_1 are the sample proportion and sample size of the first sample, and $\hat{p}_2$ and n_2 are the sample proportion and sample size of the second sample. Z is the appropriate value from the standard normal distribution for your desired confidence level. (See Chapter 3 for sample proportions and Chapter 10 [Table 10-1] for Z-values.)

To calculate a CI for the difference between two population proportions, do the following:

1. **Determine the confidence level and find the appropriate Z-value.**

   See Chapter 10 (Table 10-1).

2. **Find the sample proportion $\hat{p}_1$ for the first sample by taking the total number from the first sample that are in the category of interest and dividing by the sample size, n_1. Similarly, find $\hat{p}_2$ for the second sample.**

3. **Take the difference between the sample proportions $(\hat{p}_1 - \hat{p}_2)$.**

4. **Find $\hat{p}_1$ times $(1 - \hat{p}_1)$ and divide that by n_1. Find $\hat{p}_2$ times $(1 - \hat{p}_2)$ and divide that by n_2. Add these two results together and take the square root.**

5. **Multiply Z times the result from Step 4.**

   This is the margin of error.

6. **Take $(\hat{p}_1 - \hat{p}_2)$ plus or minus the margin of error from Step 5 to obtain the CI.**

   The lower end of the CI is $(\hat{p}_1 - \hat{p}_2)$ minus the margin of error and the upper end of the CI is $(\hat{p}_1 - \hat{p}_2)$ plus the margin of error.

While performing any calculations involving sample percentages, you must use the decimal form. After the calculations are finished, you may convert to percentages by multiplying by 100. To avoid round-off error, keep at least 2 decimal places throughout.

Suppose you work for the Las Vegas Chamber of Commerce, and you want to estimate with 95% confidence the difference between the proportion of females who have ever gone to see an Elvis impersonator and the percentage of males who have ever gone to see an Elvis impersonator, in order to help determine how you should market your entertainment offerings.

Because you want a 95% confidence interval, your Z-value is 1.96.

Suppose your random sample of 100 females includes 53 females who have seen an Elvis impersonator, so $\hat{p}_1$ is $53 \div 100 = 0.53$. Suppose also that your random sample of 110 males includes 37 males who have ever seen an Elvis impersonator, so $\hat{p}_2$ is $37 \div 110 = 0.34$.

The difference between these sample proportions (females – males) is 0.53 – 0.34 = 0.19.

Take 0.53 times (1 – 0.53) and divide that by 100 to get 0.2491 ÷ 100 = 0.0025. Then take 0.34 times (1 – 0.34) and divide that by 110 to get 0.2244 ÷ 110 = 0.0020. Add these two results to get 0.0025 + 0.0020 = 0.0045; the square root is 0.0671.

1.96 × 0.0671 gives you 0.13, or 13%, which is the margin of error.

Your 95% confidence interval for the difference between the percentage of females who have seen an Elvis impersonator and the percentage of males who have seen an Elvis impersonator is 0.19 or 19% (which you got in Step 3), plus or minus 13%. The lower end of the interval is 0.19 – 0.13 = 0.06 or 6%; the upper end is 0.19 + 0.13 = 0.32 or 32%. So you can say with 95% confidence that a higher percentage of females than males have seen an Elvis impersonator, and the difference in these percentages is somewhere between 6% and 32%, based on your sample. Now would the guys actually admit they'd ever seen an Elvis impersonator? This may create some bias in the results. (The last time I was in Vegas, I thought I really saw Elvis; he was driving a van taxi to and from the airport . . . .)

Notice that you could get a negative value for $(\hat{p}_1 - \hat{p}_2)$. For example, if you had switched the males and females, you would have gotten –0.19 for this difference. A positive difference means the first group has a larger value than the second group; a negative difference means the first group has a smaller value than the second group. You can avoid negative differences by always having the group with the larger value serve as the first group.

Part VI
Putting a Claim to the (Hypothesis) Test

The 5th Wave   By Rich Tennant

I'm mathematically dyslexic. But it's not that unusual—100 out of every 15 people are.

In this part . . .

Many statistics form the basis of claims, like "Four out of five dentists surveyed recommend this gum" or "Our diapers are 25 percent more absorbent than the leading brand." How can you tell whether the claim is true? Researchers (who know what they're doing) use what's called a hypothesis test.

In this part, you explore the basics of hypothesis tests, determining how to set them up, carry them out, and interpret the results (all the while knowing that you're trying to make a statement about an entire population based on only a sample). You also get quick references and examples for the most commonly used hypothesis tests.

Chapter 14

Claims, Tests, and Conclusions

. . . .

In This Chapter

▶ Testing other people's claims

▶ Using statistics as your evidence

▶ Weighing the evidence and making decisions

▶ Knowing that you could be wrong

. . . .

*Y*ou hear claims involving statistics all the time; the media has no shortage of them:

✔ Twenty-five percent of all women in the United State have varicose veins. (Wow, are some claims better left unsaid, or what?)

✔ Ecstasy use in teens dropped for the first time in recent years. The one-year decline ranged from about one-tenth to nearly one-third, depending on what grade they were in.

✔ A 6-month-old baby sleeps an average of 14 to 15 hours in a 24-hour period. (Yeah, right!)

✔ A name-brand ready-mix pie takes only 5 minutes to make.

Many claims involve numbers that seem to come out of thin air. Some claims make comparisons between one product or group and another. You may wonder whether such claims are valid, and you should. Not all claims are life changing (after all, what's the harm in using a soap that isn't 99.99 percent pure?) but some claims are — for example, which cancer treatment works best, which minivan is the safest, or whether certain drugs should be approved. While many claims are backed up by solid scientific (and statistically sound) research, other claims are not. In this chapter, you find out how to use statistics to determine whether a claim is actually valid and get the low-down on the process that researchers *should* be using to validate every claim they make.

Responding to Claims: Some Do's and Don'ts

In today's age of information (and big money), a great deal rides on being able to back up your claims. Companies that say their products are better than the leading brand better be able to prove it, or they could face lawsuits. Drugs that are approved by the FDA have to show strong evidence that their products actually work without producing life-threatening side effects. Manufacturers have to make sure their products are being produced according to specifications to avoid recalls, customer complaints, and loss of business.

Research can also result in claims that can mean the difference between life and death, such as which cancer treatment is best, which side effects of a type of surgery are most common, what the survival rate of a certain treatment is, and whether or not a new experimental drug increases life expectancy. The research that goes into answering these questions needs to be sound, so that the right decision (at least the most statistically informative decision) can be made. If not, researchers can lose their reputations, credibility, and funding. (And sometimes, they feel pressure to produce results, which can lead to other problems, as well.)

Knowing your options

As a consumer in this age of information, when you hear a claim being made (for example, "Our ice cream was the top choice of 80% of taste testers"), you basically have three options:

- ✔ Believe it automatically (or go the other way and reject it outright)
- ✔ Conduct your own test to verify or refute the claim
- ✔ Dig deeper for more information so you can make your own decision

Believing results without question (or rejecting them out of hand) isn't wise; the only times you may want to do this are when the source has already established a good (or bad) name with you or the result simply isn't that important (after all, you can't go around checking *every* single claim that comes your way). More on the other two options in the two following sections.

Steering clear of anecdotes

The second option for responding to a claim, the test-it-yourself approach, is one that is taken by many organizations, such as The Gallup Organization, which conducts its own polls; the Insurance Institute for Highway Safety, which crash tests and reports on the safety of vehicles; *Consumer Reports*, which tests and reports on product quality and value; and the Good Housekeeping Institute, which tests products before giving them its Seal of Approval.

The test-it-yourself approach can be effective if done correctly, with data that are based on well-designed studies that collect accurate, unbiased data (see Chapters 16 and 17 for more on study designs).

This approach is often taken in the workplace; for example, a competitor may make claims about its product that you think are untrue and should be tested. Or, you may think your product does a better job than a competitor's product, and you want to put the products to the test. Many manufacturers also do their own quality control (see Chapter 19), so they make a practice of testing their products to see whether they are within specifications.

Yet while this option is viable for groups that have the resources and the knowledge to undertake a proper study to test a claim, it can lead to misleading results if handled improperly.

One way that the media tests product claims is by sending people out into the field to check products out for themselves. This is an overused and unscientific (yet fun) method for testing a hypothesis. For example, suppose some TV show has determined that the world has to know: Does it really take five minutes to make a certain name-brand five-minute pie? Maybe it actually takes more, maybe it takes less. Statistically speaking, the variable of interest here is numerical — preparation time — and the population is all of the pies made using that name-brand recipe. The parameter of interest is the *average* preparation time for all pies made with that recipe. (A parameter is a single number that summarizes the population and is typically what the claim is about.) The claim here is that the average preparation time is 5 minutes. Their mission: Test this claim. How many pies will they use? Take a guess — it's just one!

They'd have cameras rolling, and the co-hosts would banter about how much fun it is to make the pie, how good it looks, keeping an eye on the time to prepare it (after all, they have to go to a commercial break soon). In the end, they'd report that it took them say 5.5 minutes, pretty close to the claim, but not exactly. And they'd end with a comment that using Snickers bars on top of this name-brand pie is a good candy bar choice (it is, by the way).

If these TV shows ever had a resident statistician who could give the statistical breakdown of the results (a real ratings booster, I know), I'd jump at that chance. The main idea I'd want to get across to the audience is that sample results vary (from person to person, from pie to pie) — see Chapter 9 for more on this. Measuring and understanding this variability is where the real statistics comes in. The bottom line is: To get credible, conclusive results about any claim requires real data (that is, more than a single observation). Many people don't realize that in order to properly test a claim, it takes much more than a sample size of 1 (or even 2 or 3), because of the fact that sample results vary.

You can't (or at least shouldn't) build any kind of lasting conclusions based on an anecdote, which is what a sample size of 1 really is. In statistics, a sample size of 1 doesn't make any sense. You can't measure any amount of variability with only one value (see Chapter 5 for the standard deviation formula to see what I mean). That's the trouble with most TV segments that show people testing claims by testing 1 or 2 individual products; they aren't doing a scientific test, and they send the wrong message about the way you test a hypothesis. Now while making conclusions about five-minute pies without sufficient data doesn't seem earth shattering, think about how many times hearing one person's single experience has influenced a decision you've made in your life.

Beware of any study results based on extremely small sample sizes, especially those based on a sample size of 1. For example, if a study sends an individual out to test one package of meat, examine one child's toy, or test the accuracy of one individual pharmacy filling one prescription on one particular day, steer clear. These make for interesting stories and may uncover problems to be more fully investigated, but these results alone aren't scientific, and you shouldn't make any conclusions based on them.

Digging deeper

Digging deeper to get more information is the way you want to respond to claims that are important to you. Digging deeper gives you the information you need to ask the hard questions and make an informed decision.

The biggest difference between a statistically sound test of a claim and the man-on-the-street test of a claim is that a good test uses data that have been collected in a scientific, unbiased way, based on random samples that are large enough to get accurate information. (See Chapter 2 for more on this.) Most scientific research, including medical, pharmaceutical, engineering, and government research, is based on using statistical studies to test, verify, or refute claims of various types. Being a consumer of much of this information, oftentimes just from tiny sound bytes on TV, you need to be able to know what to look for to evaluate the study, understand the results, and make your own decisions about the claims being made.

You may wonder how much protection you have as a consumer regarding claims that researchers make. The U.S. government regulates and monitors a great deal of the research and production that goes on (for example, the FDA regulates drug research and distribution, the USDA monitors food production, and so on). But some areas, such as dietary supplements (vitamins, herbal and mineral supplements, and so on), aren't as rigorously regulated.

As a consumer of all the results thrown at you in today's society, you need to be armed with information to make good decisions. A good first step is to contact the researcher (or the journalist) to see whether any scientific studies back up his or her claim. If he or she says yes, ask whether you can see the descriptions and results of those studies, and then evaluate that information critically (see Chapters 16 and 17 for more on this).

Doing a Hypothesis Test

A *hypothesis test* is a statistical procedure that's designed to test a claim. Typically, the claim is being made about a population parameter (one number that characterizes the entire population). Because parameters tend to be unknown quantities, everyone wants to make claims about what their values may be. For example, the claim that 25% (or 0.25) of all women have varicose veins is a claim about the proportion (that's the *parameter*) of all women (that's the *population*) who have varicose veins (that's the *variable*, having or not having varicose veins).

Do you think that anyone actually knows for certain that the percentage of all women who have varicose veins is exactly 25? No; they're making a claim, not stating a fact. Watch out for statements like this.

Defining what you're testing

To get more specific, the varicose vein claim is that the parameter, the population proportion (*p*), is equal to 0.25. (This claim is called the *null hypothesis.*) If you're out to test this claim, you're questioning the claim and have a hypothesis of your own (called the *research hypothesis,* or *alternative hypothesis*). You may hypothesize, for example, that the actual proportion of women who have varicose veins is lower than 0.25, based on your observations. Or, you may hypothesize that due to the popularity of high heeled shoes, the proportion may be higher than 0.25. Or, if you're simply questioning whether the actual proportion is 0.25, your alternative hypothesis is, "No, it isn't 0.25."

In addition to testing hypotheses about categorical variables (having or not having varicose veins is a categorical variable), you can also test hypotheses about numerical variables, such as the average commuting time for people working in Los Angeles or their average household income. In these cases, the parameter of interest is the population average or mean (denoted μ). Again, the claim is that this parameter is equal to a certain value, versus some alternative.

Hypotheses can be tested about more than one single population parameter, too. For example, you may want to compare average household incomes or commuting times of people from two or more major cities. Or you may want to see whether a link exists between commuting time and income. All of these questions can be answered using hypothesis tests; while the details differ for each situation, the general ideas are the same. I go over the one-sample case for means and proportions (large samples) in this chapter; Chapter 15 provides the particulars of many commonly used hypothesis tests.

Setting up the hypotheses

Every hypothesis test contains two hypotheses. The first hypothesis is called the *null hypothesis,* denoted H_o. The null hypothesis always states that the population parameter is *equal* to the claimed value. For example, if the claim is that the average time to make a name-brand ready-mix pie is five minutes, the statistical shorthand notation for the null hypothesis in this case would be as follows: H_o: $\mu = 5$.

What's the alternative?

Before actually conducting a hypothesis test, you have to put two possible hypotheses on the table — the null hypothesis is one of them. But, if the null hypothesis is found not to be true, what's your alternative going to be? Actually, three possibilities exist for the second (or alternative) hypothesis, denoted H_a. Here they are, along with their shorthand notations in the context of the example:

- The population parameter is *not equal to* the claimed value (H_a: $\mu \neq 5$).

- The population parameter is *greater than* the claimed value (H_a: $\mu > 5$).

- The population parameter is *less than* the claimed value (H_a: $\mu < 5$).

Which alternative hypothesis you choose in setting up your hypothesis test depends on what you're interested in concluding, should you have enough evidence to refute the null hypothesis (the claim).

For example, if you want to test whether a company is correct in claiming its pie takes 5 minutes to make and you also want to know whether the actual average time is more or less than that, you use the not-equal-to alternative. Your hypotheses for that test would be H_o: $\mu = 5$ versus H_a: $\mu \neq 5$.

If you only want to see whether the time turns out to be greater than what the company claims (that is, the company is falsely advertising its prep time), you use the greater-than alternative, and your two hypotheses are H_o: $\mu = 5$ versus H_a: $\mu > 5$.

Finally, say you work for the company marketing the pie, and you think the pie can be made in less than 5 minutes (and could be marketed by the company as such). The less-than alternative is the one you want, and your two hypotheses would be H_o: $\mu = 5$ versus H_a: $\mu < 5$.

Knowing which hypothesis is which

How do you know which hypothesis to put in H_o and which one to put in H_a? Typically, the null hypothesis says that nothing new is happening; the previous result is the same now as it was before, or the groups have the same average (their difference is equal to zero). In general, you assume that people's claims are true until proven otherwise.

Hypothesis tests are similar to jury trials, in a sense. In a jury trial, H_o is similar to the not-guilty verdict, and H_a is the guilty verdict. You assume in a jury trial that the defendant isn't guilty unless the prosecution can show beyond a reasonable doubt that he or she *is* guilty. If the jury says the evidence is beyond a reasonable doubt, they reject H_o, not guilty, in favor of H_a, guilty.

In general, when hypothesis testing, you set up H_o and H_a so that you believe H_o is true unless your evidence (your data and statistics) shows you otherwise. And in that case, where you have sufficient evidence against H_o, you reject H_o in favor of H_a. The burden of proof is on the researcher to show sufficient evidence against H_o before it's rejected. (That's why H_a is often called the research hypothesis, because H_a is the hypothesis that the researcher is most interested in showing.) If H_o is rejected in favor of H_a, the researcher can say he or she has found a *statistically significant* result; that is, the results refute the previous claim, and something different or new is happening.

In many cases, people set up hypothesis tests because they're out to show that H_o isn't true, supporting the alternative hypothesis. (The mentality is, why do research just to show that something has stayed the same?) The results that you hear about in the media are generally the ones that are able to show H_o isn't true; this is what makes news. In many cases that's a good thing, though, because researchers and manufacturers have to stay on their toes to avoid negative publicity surrounding a product recall, a lawsuit, or a government investigation. That's because if one of their claims (H_o) is rejected by someone conducting an independent hypothesis test, the researchers or manufacturers are being judged as guilty of false advertising or false claims, which is not good.

Gathering the evidence: The sample

After you've set up the hypotheses, the next step is to collect your evidence and determine whether your evidence corroborates the claim made in H_o. Remember, the claim is made about the population, but you can't test the whole population; the best you can usually do is take a sample. As with any other situation in which statistics are being collected, the quality of the data is extremely critical. (See Chapter 2 for lots of examples of statistics that have gone wrong.)

Good data start with a good sample. The two main issues to consider when selecting your sample are avoiding bias and being accurate. To avoid bias, take a random sample (meaning everyone in the population must have an equal chance of being chosen) and choose a large enough sample size so that the results will be accurate. (See Chapter 3.)

Compiling the evidence: The statistics

After you select your sample, the appropriate number-crunching takes place. Your null hypothesis makes a statement about what the population parameter is (for example, the proportion of all women who have varicose veins or the average miles per gallon of a U.S.-built light truck). In statistical jargon, the data you collect measure that variable of interest, and the statistics that you calculate will include the sample statistic that most closely estimates the population parameter. In other words, if you're testing a claim about the proportion of women with varicose veins, you need to calculate the proportion of women in your sample who have varicose veins. If you're testing a claim about the average miles per gallon of a U.S.-built light truck, your statistic should be the average miles per gallon of the light trucks in your sample. (See Chapter 5 for all the information you need on calculating statistics.)

Standardizing the evidence: The test statistic

After you have your sample statistic, you may think you're done with the analysis part and are ready to make your conclusions — but you're not. The problem is, you have no way to put your results into any kind of perspective just by looking at them in their regular units. That's because you know that your results are based only on a sample and that sample results are going to vary. That variation needs to be taken into account, or your conclusions could be completely wrong. (How much do sample results vary? Sample variation is measured by the standard error; see Chapter 9 for more on this.)

Suppose the claim is that the percentage of all women with varicose veins is 25 percent, and your sample of 100 women had 20 percent with varicose veins. The standard error for your sample percentage is 4 percent (according to formulas in Chapter 9), which means that your results are expected to vary by about twice that, or about 8 percent, according to the empirical rule (see Chapter 10). So a difference of 5 percent between the claim and your sample result (25% – 20% = 5%) isn't that much, in these terms. This represents a distance of less than 2 standard errors away from the claim. Therefore, you accept the claim, H_o, because your data can't refute it.

However, suppose your sample percentage was based on a sample of 1,000 women, not 100. This decreases the amount by which you expect your results to vary, because you have more information. The standard error becomes 0.012 or 1.2 percent, and the margin of error is twice that, or 2.4 percent on either side. Now a difference of 5 percent between your sample result (20 percent) and the claim (25 percent) is a more meaningful difference; it's *way* more than 2 standard errors away from the claim. Your results being based on 1,000 people shouldn't vary that much from the claim, so what should you conclude? The claim (H_o) is concluded to be false, because your data don't support it.

The number of standard errors that a statistic lies above or below the mean is called a *standard score* (see Chapter 8). In order to interpret your statistic, you need to convert it from original units to a standard score. When finding a standard score, you take your statistic, subtract the mean, and divide the result by the standard error. In the case of hypothesis tests, you use the value in H_o as the mean. (That's because you assume H_o is true, unless you have enough evidence against it.) This standardized version of your statistic is called a *test statistic,* and it's the main component of a hypothesis test. (Chapter 15 contains the formulas for the most common hypothesis tests.)

The general procedure for converting a statistic to a test statistic (standard score) in the case of means/proportions:

1. **Take your statistic minus the claimed value (given by H_o).**

2. **Divide by the standard error of the statistic (see Chapters 9 and 10).**

Your test statistic represents the distance between your actual sample results and the claimed population value, in terms of number of standard errors. In the case of a single population mean or proportion, you know that these standardized distances should have a standard normal distribution if your sample size is large enough (see Chapters 8 and 9). So, to interpret your test statistic in these cases, you can see where it stands on the standard normal distribution (Z-distribution).

Although you never expect a sample statistic to be exactly the same as the population value, you expect it to be close if H_o is true. That means, if you see that the distance between the claim and the sample statistic is small, in terms of standard errors, your sample isn't far from the claim and your data are telling you to stick with H_o. As that distance becomes larger and larger, however, your data are showing less and less support for H_o. At some point, you should reject H_o based on your evidence, and choose H_a. At what point does this happen? The next section addresses that issue.

Weighing the Evidence and Making Decisions: P-Values

To test whether the claim is true, you're looking at your test statistic taken from your sample, and seeing whether it supports the claim. And how do you determine that? For most cases, by looking at where your test statistic ends up on the standard normal distribution (Z-distribution) — see Chapter 9. The Z-distribution has a mean of 0 and a standard deviation of 1. If your test statistic is close to 0, or at least within that range where most of the results should fall, then you say yes, the claim (H_o) is probably true, and the sample results verify it. If your test statistic is out in the tails of the standard normal distribution, then you say no, the chance of my sample results ending up this far out on the distribution is too small; my sample results don't verify the claim (H_o).

But how far is "too far" from 0? As long as you have a large enough sample size, you know that your test statistic falls somewhere on a standard normal distribution, according to the central limit theorem (see Chapter 10). If the null hypothesis is true, most (about 95%) of the samples will result in test statistics that lie roughly within 2 standard errors of the claim. If H_a is the not-equal-to alternative, any test statistic outside this range will result in H_o being rejected (see Figure 14-1).

Note that if the alternative hypothesis is the less-than alternative, you reject H_o only if the test statistic falls in the left tail of the distribution. Similarly, if H_a is the greater-than alternative, you reject H_o only if the test statistic falls in the right tail.

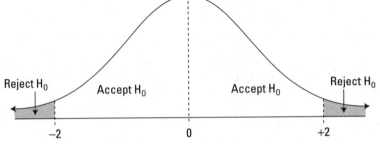

Figure 14-1:
Test statistics and your decision.

Reject H_0 Accept H_0 Accept H_0 Reject H_0

−2   0   +2

P-value basics

You can be more specific about your conclusion by noting exactly how far out on the standard normal distribution the test statistic falls, so everyone knows where the result stands and what that means in terms of how strong the evidence is against the claim. You do this by looking up the test statistic on the standard normal distribution (Z-distribution) and finding the probability of being at that value or beyond it (in the same direction) by using Table 8-1 (see Chapter 8). The *p*-value measures how likely it was that you would have gotten your sample results if the null hypothesis were true. The farther out your test statistic is on the tails of the standard normal distribution, the smaller the *p*-value will be, and the more evidence you have against the null hypothesis being true.

All *p*-values are probabilities between 0 and 1.

To find the p-value for your test statistic (means/porportions, large samples):

1. **Look up the location of your test statistic on the standard normal distribution (see Table 8-1 in Chapter 8).**

2. **Find the percentage chance of being at or beyond that value in the same direction:**

    a. If H_a contains a less-than alternative, find the percentile from Table 8-1 in Chapter 8 that corresponds to your test statistic.

    b. If H_a contains a greater-than alternative, find the percentile from Table 8-1 in Chapter 8 that corresponds to your test statistic, and then take 100% minus that. (You want the percentage to the right of your test statistic in this case, and percentiles give you the percentage to the left. See Chapter 5.)

3. **Double this percentage if (and only if) H_a is the not-equal-to alternative.**

    This accounts for both the less-than and the greater-than possibilities.

4. **Change the percentage to a probability by dividing by 100 or moving the decimal point two places to the left.**

To interpret a p-value:

✔ For small p-values (generally less than 0.05), reject H_o. Your data don't support H_o, and your evidence is beyond a reasonable doubt.

✔ For large p-values (generally greater than 0.05), you can't reject H_o. You don't have enough evidence against it.

✔ If your p-value is on or close to the borderline between accepting and rejecting, your results are marginal. (They could go either way.)

Generally, statisticians stay with H_o unless the evidence is beyond a reasonable doubt, just like in a courtroom. What probability reflects that cutoff point? It can be rather arbitrary (the term "small p-value" can mean something different to each person). For most statisticians, if the p-value is less than 0.05 given the data they collect, they'll reject H_o, and choose H_a. Some people may have stricter cutoffs, such as 0.01, requiring more evidence before rejecting H_o. Each reader makes his/her own decision. That's why researchers need to report p-values, rather than just their decisions, so that people can come to their own conclusions based on their own internal cutoff points. For example, if your p-value is 0.026 when testing H_o: $p = 0.25$ versus H_a: $p < 0.25$ in the varicose veins example, a reader with a personal cutoff point of 0.05 would conclude that H_o is false, because the p-value (of 0.026) is less than 0.05. However, a reader with a personal cutoff of 0.01 would not have enough evidence (based on your sample) to reject H_o, because the p-value of 0.026 is greater than 0.01.

Caution: Interpretations will vary!

Some people do like to set a cutoff probability before doing a hypothesis test; this is called an *alpha level* (α). Typical values for α are 0.05 or 0.01. Here's how they interpret their results in that case:

- ✔ If the *p*-value is greater than or equal to α, accept H_o.
- ✔ If the *p*-value is less than α, reject H_o.
- ✔ *P*-values on the borderline (very close to α) are treated as marginal results.

Other people don't set a predetermined cutoff; they just report the *p*-value and interpret their results by looking at the size of the *p*-value. Generally speaking,

- ✔ If the *p*-value is less than 0.01 (very small), the results are considered highly statistically significant — reject H_o.
- ✔ If the *p*-value is between 0.05 and 0.01 (but not close to 0.05), the results are considered statistically significant — reject H_o.
- ✔ If the *p*-value is close to 0.05, the results are considered marginally significant — decision could go either way.
- ✔ If the *p*-value is greater than (but not close to) 0.05, the results are considered non-significant — accept H_o.

When you hear about a result that has been found to be statistically significant, ask for the *p*-value and make your own decision. Cutoff points and resulting decisions vary from researcher to researcher.

Knowing That You Could Be Wrong: Errors in Testing

After you make a decision to either reject H_o or accept H_o, the next step is living with the consequences, in terms of how people respond to your decision.

- ✔ If you conclude that a claim isn't true but it actually *is* true, will that result in a lawsuit, a fine, unnecessary changes in the product, or consumer boycotts that shouldn't have happened?
- ✔ If you conclude that a claim is true but it actually isn't, what happens then? Will products continue to be made in the same way as they are now? Will no new law be made, no new action taken, because you showed that nothing was wrong?

Every hypothesis test decision has impact; otherwise, why do the tests?

So, a consequence can result from any decision: You could be wrong! The *X-Files* motto applies here: "The truth is out there." But the thing is, you don't know what the truth is; that's why you did the hypothesis test in the first place.

Making a false alarm: Type-1 errors

Suppose a company claims that its average package delivery time is 2 days, and a consumer group tests this hypothesis and concludes that the claim is false: They believe that the average delivery time is actually more than 2 days. This is a big deal. If the group can stand by its statistics, it has done well to inform the public about the false advertising issue. But what if the group is wrong? Even if the study is based on a good design, collects good data, and makes the right analysis, the group can still be wrong.

Why? Because its conclusions were based on a sample of packages, not on the entire population. And Chapter 9 tells you, sample results vary from sample to sample. If your test statistic falls on the tail of the standard normal distribution, these results are unusual, if the claim is true, because you expect them to be much closer to the middle of the standard normal distribution (Z-distribution). Just because the results from a sample are unusual, however, doesn't mean they're impossible. A *p*-value of 0.04 means that the chance of getting your particular test statistic (out on the tail of the standard normal distribution), even if the claim is true, is 4% (less than 5%). That's why you reject H_o in this case, because that chance is so small. But a chance is a chance!

Perhaps your sample, while collected randomly, just happens to be one of those atypical samples whose result ended up far out on the distribution. So H_o could be true, but your results lead you to a different conclusion. How often does that happen? Five percent of the time (or whatever your given cutoff probability is for rejecting H_o).

Rejecting H_o when you shouldn't is called a *type-1 error*. I don't really like this name, because it seems so nondescript. I prefer to call a type-1 error a *false alarm*. In the case of the packages, if the consumer group made a type-1 error when it rejected the company's claim, they created a false alarm. What's the result? A very angry delivery company, I guarantee that!

Missing a detection: Type-2 errors

On the other hand, suppose the company really wasn't delivering on its claim. Who's to say that the consumer group's sample will detect it? If the actual delivery time is 2.1 days instead of 2 days, the difference would be pretty hard to detect. If the actual delivery time is 3 days, a fairly small sample would show that something's up. The issue lies with those in-between values, like 2.5 days. If H_o is indeed false, you want to find out about it and reject H_o. Not rejecting H_o when you should have is called a *type-2 error*. I like to call it a *missed detection*.

Sample size is the key to being able to detect situations where H_o is false and to avoiding type-2 errors. The more information you have, the less variable your results will be (see Chapter 8) and the more ability you have to zoom in on detecting problems that exist with a claim.

This ability to detect when H_o is truly false is called the *power* of a test. Power is a pretty complicated issue, but what's important for you to know is that the higher the sample size, the more powerful a test is. A powerful test has a small chance for a type-2 error.

Take any statistically significant results with a grain of salt, no matter how well the study was conducted. Whatever decision was made, that decision could be wrong. If the study is set up right, however (see Chapter 16 for surveys and Chapter 17 for experiments) that chance should be fairly small.

Statisticians recommend two preventative measures to minimize the chances of a type-1 or type-2 error:

- ✔ Set a low cutoff probability for rejecting H_o (like 5 percent or 1 percent) to reduce the chance of false alarms (minimizing type-1 errors).

- ✔ Select a large sample size to ensure that any differences or departures that really exist won't be missed (minimizing type-2 errors).

Drawing conclusions about their conclusions

Even if you never conduct a hypothesis test of your own, just knowing how they are supposed to be done can sharpen your critiquing skills. After the test is finished, the next step for researchers is to publish the results and offer press releases to the media indicating what they found. This is another place where you need to be watchful. While many researchers are good about stating their results carefully and pointing out the limitations of their data, others take a bit more liberty with their conclusions (whether they intend to do that or not is a separate issue).

Walking through a Hypothesis Test: The Big Picture

Every hypothesis test contains a series of steps and procedures. This section gives you a general breakdown of what's involved. See Chapter 15 for details on the most commonly used hypothesis tests, including tests that examine a claim about a single population parameter, as well as those that compare two populations.

Reviewing the general steps for a hypothesis test (one means/proportions, large samples)

Here's a boiled-down summary of the calculations involved in doing a hypothesis test. (Particular formulas needed to find test statistics for any of the most common hypothesis tests are provided in Chapter 15.)

1. **Set up the null and alternative hypotheses:**

    a. The null hypothesis, H_o, says that the population parameter is equal to some claimed number.

    b. Three possible alternative hypotheses exist; choose the one that's most relevant in the case where the data *don't* support H_o.

    i. H_a: The population parameter is *not equal (≠) to* the claimed number.

    ii. H_a: The population parameter is *less than (<)* the claimed number.

    iii. H_a: The population parameter is *greater than (>)* the claimed number.

2. **Take a random sample of individuals from the population and calculate the sample statistic.**

    This gives your best estimate of the population parameter (see Chapter 4).

3. **Convert the sample statistic to a test statistic by changing it to a standard score (all formulas for test statistics are provided in Chapter 15):**

    a. Take your sample statistic minus the number in the null hypothesis. This is the distance between the claim and your results.

    b. Divide that distance by the standard error of your statistic (see Chapter 10 for more on standard error). This changes the distance to standard units.

4. **Find the *p*-value for your test statistic.**

   a. Find the percentage chance of being at or beyond that value in the same direction:

      i. If H_a contains a less-than alternative, find the percentile from Table 8-1 in Chapter 8 that corresponds to your test statistic.

      ii. If H_a contains a greater-than alternative, find the percentile from Table 8-1 (see Chapter 8) that corresponds to your test statistic, and then take 100% minus that percentile. (This gives you the percentage to the right of your test statistic.)

   b. Double this percentage if (and only if) H_a is the not-equal-to alternative.

   c. Change the percentage to a probability by dividing by 100 or by moving the decimal point two places to the left. This is your *p*-value.

5. **Examine your *p*-value and make your decision.**

   a. Smaller *p*-values show more evidence against H_o. Conclude that H_o is false (in other words, reject the claim).

   b. Larger *p*-values show more evidence for H_o. Conclude that you can't reject H_o. Your sample supports the claim.

      What's the cutoff point between having or not having enough support for H_o? Most people find 0.05 to be a good cutoff point for accepting or rejecting H_o; *p*-values less than 0.05 show reasonable doubt that H_o is true. Your cutoff point is called the alpha (α) level.

In a case where two populations are being compared, most researchers are interested in comparing the groups according to some parameter, such as the average weight of males versus females, or the proportion of women who oppose an issue compared to the proportion of men. In this case, the hypotheses are set up so you're looking at the difference between the averages or proportions, and the null hypothesis is that the difference is zero (the groups have the same means or proportions). Chapter 15 gives formulas and examples for these hypothesis tests for both the large and small sample size cases.

Dealing with other hypothesis tests

Many types of hypothesis tests are done in the world of research. The most common ones have been included in Chapter 15 (along with easy-to-use formulas, step-by-step explanations, and examples). But so many types of tests exist and their results come to you on an everyday basis — many of them in sound bytes, press releases, evening news broadcasts, and on the Internet.

While the hypothesis tests that researchers use can be quite varied, the main ideas (such as *p*-values and how to interpret those results) are the same.

The most important element that all hypothesis tests have in common is the *p*-value. All *p*-values have the same interpretation, no matter what test is done. So anywhere you see a *p*-value, you will know that a small *p*-value means the researcher found a "statistically significant" result, which means the null hypothesis was rejected.

You also know, regardless of which hypothesis test someone used, that any conclusions that are made are subject to the process of data collection and analysis being done correctly. Even then, under the best of situations, the data could still be unrepresentative just by chance, or the truth could have been too hard to detect, and the wrong decision could be made. But that's part of what makes statistics such a fun subject — you never know if what you're doing is correct, but you always know that what you're doing is right; does that make sense?

Handling smaller samples: The t-distribution

For means/porportions, in the case where the sample size is small (and by small, I mean dropping below 30 or so), you have less information on which to base your conclusions. Another drawback is that you can't rely on the standard normal distribution (Z-distribution) to compare your test statistic, because the central limit theorem hasn't kicked in yet. (The central limit theorem requires sample sizes that are large enough for the results to average out to a bell-shaped curve; see Chapter 8 for more on this.) You already know you should disregard results that are based on very small sample sizes (especially those with a sample size of 1). So, what do you do in those in-between situations, in which the sample size isn't small enough to disregard and isn't large enough to use the standard normal distribution to weigh your evidence? You use a different distribution, called a *t-distribution.* (You may have heard of the term *t-test* before, in terms of hypothesis testing. This is where that term comes from.)

The t-distribution is basically a shorter, fatter version of the standard normal distribution (Z-distribution). The idea is, you should have to pay a penalty for having less information, and that penalty is a distribution that has fatter tails. To make a touchdown (getting into that magic 5% range where H_0 is rejected) with a smaller sample size is going to mean having to go farther out, proving yourself more, and having stronger evidence than you normally would if you had a larger sample size. Figure 14-2 compares the standard normal distribution (Z-distribution) to a t-distribution.

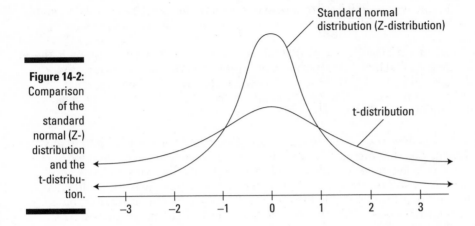

Figure 14-2:
Comparison
of the
standard
normal (Z-)
distribution
and the
t-distribu-
tion.

Each sample size has its own t-distribution. That's because the penalty
for having a smaller sample size, like 5, is greater than the penalty for having
a larger sample size, like 10 or 20. Smaller sample sizes have shorter, fatter
t-distributions than the larger sample sizes. And as you may expect, the larger
the sample size is, the more the t-distribution looks like a standard normal
distribution (Z-distribution); and the point where they become very similar
(similar enough for jazz or government work) is about the point where the
sample size is 30. Figure 14-3 shows what different t-distributions look like for
different sample sizes and how they all compare to the standard normal distribution (Z-distribution).

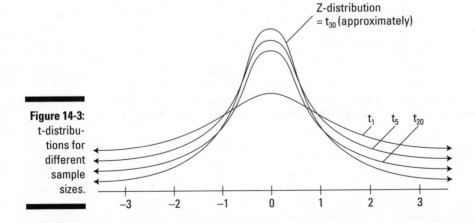

Figure 14-3:
t-distribu-
tions for
different
sample
sizes.

Each t-distribution is distinguished by something statisticians call *degrees of
freedom.* (Why they call it that is something that goes beyond this book.) When
you're testing one population's mean and the sample size is n, the degrees of
freedom for the corresponding t-distribution is $n - 1$. So, for example, if your
sample size is 10, you use a t-distribution with $10 - 1$ or 9 degrees of freedom,

denoted t_9, rather than a Z-distribution, to look up your test statistic. (For any test that uses the t-distribution, the degrees of freedom will be given in terms of a formula involving the sample sizes. See Chapter 15 for details.)

The t-distribution makes you pay a penalty for having a small sample size. What's the penalty? A larger *p*-value than one that the standard normal distribution would have given you for the same test statistic. That's because of the fatter tails on the t-distribution; a test statistic far out on the leaner Z-distribution has little area beyond it. But that same test statistic out on the fatter t-distribution has more fat (or area) beyond it, and that's exactly what the *p*-value represents. A bigger *p*-value means less chance of rejecting H_o. Having less data should create a higher burden of proof , so *p*-values do work the way you'd expect them to, after you figure out what you expect them to do!

Because each sample size would have to have its own t-distribution with its own t-table to find *p*-values, statisticians have come up with one abbreviated table that you can use to get a general feeling for your results (see Table 14-2). Computers can also give you a precise *p*-value for any sample size.

Table 14-2		t-Distribution			

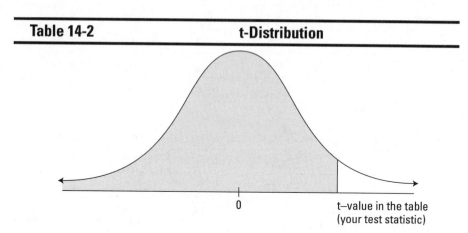

Degrees of Freedom	90th Percentile	95th Percentile	97.5th Percentile	98th Percentile	99th Percentile
1	3.078	6.314	12.706	31.821	63.657
2	1.886	2.920	4.303	6.965	9.925
3	1.638	2.353	3.182	4.541	5.841
4	1.533	2.132	2.776	3.747	4.604
5	1.476	2.015	2.571	3.365	4.032
6	1.440	1.943	2.447	3.143	3.707

(continued)

Table 14-2 (continued)

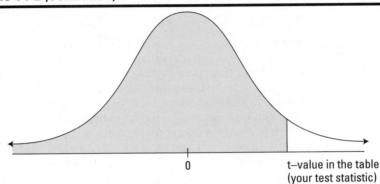

0   t–value in the table
(your test statistic)

Degrees of Freedom	90th Percentile	95th Percentile	97.5th Percentile	98th Percentile	99th Percentile
7	1.415	1.895	2.365	2.998	3.499
8	1.397	1.860	2.306	2.896	3.355
9	1.383	1.833	2.262	2.821	3.250
10	1.372	1.812	2.228	2.764	3.169
11	1.363	1.796	2.201	2.718	3.106
12	1.356	1.782	2.179	2.681	3.055
13	1.350	1.771	2.160	2.650	3.012
14	1.345	1.761	2.145	2.624	2.977
15	1.341	1.753	2.131	2.602	2.947
16	1.337	1.746	2.120	2.583	2.921
17	1.333	1.740	2.110	2.567	2.898
18	1.330	1.734	2.101	2.552	2.878
19	1.328	1.729	2.093	2.539	2.861
20	1.325	1.725	2.086	2.528	2.845
21	1.323	1.721	2.080	2.518	2.831
22	1.321	1.717	2.074	2.508	2.819
23	1.319	1.714	2.069	2.500	2.807
24	1.318	1.711	2.064	2.492	2.797

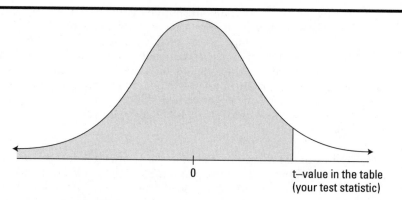

t–value in the table
(your test statistic)

Degrees of Freedom	90th Percentile	95th Percentile	97.5th Percentile	98th Percentile	99th Percentile
25	1.316	1.708	2.060	2.485	2.787
26	1.315	1.706	2.056	2.479	2.779
27	1.314	1.703	2.052	2.473	2.771
28	1.313	1.701	2.048	2.467	2.763
29	1.311	1.699	2.045	2.462	2.756
30	1.310	1.697	2.042	2.457	2.750
40	1.303	1.684	2.021	2.423	2.704
60	1.296	1.671	2.000	2.390	2.660
Z-values	1.282	1.645	1.960	2.326	2.576

Suppose your sample size is 10, your test statistic (referred to as the *t-value*) is 2.5, and your alternative hypothesis, H_a, is the greater-than alternative. Because the sample size is 10, you use the t-distribution with $10 - 1 = 9$ degrees of freedom to calculate your *p*-value. This means you'll be looking at the row in the t-table (Table 14-2) that has a 9 in the Degrees of Freedom column. Your test statistic (2.5) falls between two values: 2.262 (the 97.5th percentile) and 2.821 (the 98th percentile).

What's the *p*-value? Somewhere between $100\% - 97.5\% = 2.5\% = 0.025$ and $100\% - 98\% = 2\% = 0.02$. (Keep in mind that with a greater-than alternative, you need 100% minus the percentile.) You don't know exactly what the *p*-value is, but because 2% and 2.5 % are both less than the typical cutoff of 5%, you reject H_o.

Note that for a less-than alternative hypothesis, your test statistic would be a negative number (to the left of 0 on the t-distribution). In this case, you want to find the percentage below, or to the left of, your test statistic to get your *p*-value. Yet negative test statistics don't appear on Table 14-2. Not to worry! The percentage to the left (below) a negative t-value is the same as the percentage to the right (above) the positive t-value, due to symmetry. So, to find the *p*-value for your negative test statistic, look up the positive version of your test statistic on Table 14-2, find the corresponding percentile, and take 100% minus that.

For example, if your test statistic is –2.5 with 9 degrees of freedom, look up +2.5 on Table 14-2, and you find that it falls between the 97.5th and 98th percentiles. Taking 100% minus these amounts, your *p*-value is somewhere between 2% and 2.5%. (Note that this approach for negative numbers is different from how these situations are handled in Table 8-1 in Chapter 8. But that table was set up differently.)

If your alternative hypothesis (H_a) has the not-equal-to alternative, double the percentage that you get.

For all types of hypotheses (greater-than, less-than, and not-equal-to), change the percentage to a probability by dividing by 100 or moving the decimal point two places to the left.

The t-table (Table 14-2) doesn't include all possible test statistics on it, so simply choose the one test statistic that's closest to yours, look at the column it's in, and find the corresponding percentile. Then figure your *p*-value.

The last line of Table 14-2 shows the corresponding values from the standard normal distribution (Z-distribution) for the given percentiles. Notice that as the degrees of freedom of the t-distribution increase (as you move down any given column), the numbers get closer and closer to that last row of the table. That confirms what you already know: As the sample size increases, the t- and the Z-distributions are more and more alike.

Chapter 15

Commonly Used Hypothesis Tests: Formulas and Examples

· ·

In This Chapter

▶ Breaking down some of the most commonly used hypothesis tests

▶ Calculating their test statistics

▶ Using the results to make informed decisions

· ·

*W*hether in product advertisements or media blitzes on recent medical breakthroughs, you often run across claims made about one or more populations. For example, "We promise to deliver our packages in two days or less" or, "Two recent studies show that a high-fiber diet may reduce your risk of colon cancer by 20%." Whenever someone makes a claim (also called a *null hypothesis*) about a population (for example, that the average amount of time people spend commuting to and from work is 6 hours per week, or that the percentage of people in the United States who like reality TV is 30%), you can test the claim by doing what statisticians call a *hypothesis test*. You can also use a hypothesis test to compare two populations (for example, the mean commuting time for people working first shift compared to people working second shift, or the proportion of women compared to men who have cellphones). See Chapter 14 for background information on the general ideas behind hypothesis tests.

A hypothesis test involves setting up your *hypotheses* (a claim and its alternative), selecting a sample (or samples), collecting data, calculating the relevant statistics, and using those statistics to decide whether the claim is true. What you're really doing is comparing your sample statistic to the claimed population parameter and seeing how close they are to each other. For example, if the average commuting time for a sample of 1,000 workers is 5.2 hours, 5.2 is the sample statistic. If the claim is that the average commuting time for the population of *all* workers is 6 hours per week, 6 is the claimed population parameter, in this case, the population mean. The closer the sample statistic is to the claimed value of the population parameter, the more you can believe that the claim is valid. Yet the big question is, "How close is close enough?"

In this chapter, I outline the formulas used for some of the most common hypothesis tests, explain the necessary calculations, and walk you through some examples.

Testing One Population Mean

This test is used when the variable is numerical (for example, age, income, time, and so on) and only one population or group is being studied (for example, all U.S. households or all college students). For example, Dr. Ruth says that the average time that working mothers spend talking to their children is 11 minutes per day, on average. (For dads, the claim is 8 minutes.) The variable, time, is numerical, and the population is all working mothers.

The null hypothesis is that the population mean, μ, is equal to a certain claimed value, μ_o. The notation for the null hypothesis is $H_o : \mu = \mu_o$. So, the null hypothesis in the Dr. Ruth example is $H_o: \mu = 11$ minutes, and μ_o is 11 here. Note that μ represents the average number of minutes per day that all working mothers spend talking to their children, on average. The alternative hypothesis, H_a, is either $\mu > \mu_o, \mu < \mu_o$, or $\mu \neq \mu_o$. In this example, the three possibilities for H_a would be: $\mu > 11$, $\mu < 11$, or $\mu \neq 11$. (See Chapter 14 for more on alternative hypotheses.) If you suspect that the average time working mothers spend talking with their kids is more than 11 minutes, your alternative hypothesis would be $H_a: \mu > 11$.

The formula for the test statistic for one population mean is $\dfrac{\bar{x} - \mu_o}{s / \sqrt{n}}$. To calculate it, do the following:

1. **Calculate the sample mean, $\bar{x}$, and the sample standard deviation, s. Let n represent the sample size.**

   See Chapter 4 for calculations of the mean and standard deviation.

2. **Find $\bar{x}$ minus μ_o.**

3. **Calculate the standard error: $s / \sqrt{n}$. Save your answer.**

4. **Divide your result from Step 2 by the standard error found in Step 3.**

   For the Dr. Ruth example, suppose a random sample of 100 working mothers spend an average of 11.5 minutes per day talking with their children, with a standard deviation of 2.3 minutes. That means $\bar{x}$ is 11.5, $n = 100$, and $s = 2.3$.

   Take $11.5 - 11 = +0.5$.

   Take 2.3 divided by the square root of 100 (which is 10) to get 0.23 for the standard error.

   Divide $+0.5$ by 0.23, to get 2.17 (rounded to 2.2). That's your test statistic.

This means your sample mean is 2.2 standard errors above the claimed population mean. Would these sample results be unusual if the claim (H_o: μ = 11 minutes) were true? To decide whether your test statistic supports H_o, calculate the p-value. To calculate the p-value, look up your test statistic (in this case 2.2) on the standard normal distribution (Z-distribution) — see Table 8-1 in Chapter 8 — and take 100% minus the percentile shown, because your H_a is a greater-than hypothesis. In this case, the percentage would be 100% – 98.61% = 1.39%. So, the p-value (dividing by 100) would be 0.0139. (See Chapter 14 for more on p-value calculations.) This p-value of 0.0139 (1.39%) is quite a bit less than 0.05 (5%). That means your sample results are unusual if the claim (of 11 minutes) is true. So, reject the claim (μ = 11 minutes) by rejecting H_o, and then accept H_a (μ > 11 minutes).

Your conclusion: According to this (hypothetical) sample, Dr. Ruth's claim of 11 minutes is a bit low; the actual average is greater than 11 minutes per day. See Chapter 14 for more on hypothesis test calculations and conclusions.

If the sample size, n, were less than 30 here, you would look up your test statistic on the t-distribution, rather than the standard normal distribution (Z-distribution). See Table 14-2 in Chapter 14 for more on this. For information on how to calculate p-values for the less-than or not-equal-to alternatives, also see Chapter 14.

Testing One Population Proportion

This test is used when the variable is categorical (for example, gender, political party, support/oppose, and so on) and only one population or group is being studied (for example, all U.S. citizens or all registered voters). The test is looking at the proportion (p) of individuals in the population who have a certain characteristic, for example, the proportion of people who carry cellphones. The null hypothesis is H_o: $p = p_0$, where p_0 is a certain claimed value. For example, if the claim is that 20% of people carry cellphones, p_0 is 0.20. The alternative hypothesis is one of the following: $p > p_0$, $p < p_0$, or $p \neq p_0$. (See Chapter 14 for more on alternative hypotheses.)

The formula for the test statistic for a single proportion is $\dfrac{\hat{p} - p_o}{\sqrt{\dfrac{p_o(1 - p_o)}{n}}}$. To calculate it, do the following:

1. **Calculate the sample proportion, $\hat{p}$, by taking the number of people in the sample who have the characteristic of interest (for example, the number of people in the sample carrying cellphones) and dividing that by n, the sample size.**

2. **Take $\hat{p}$ minus p_0.**

3. **Calculate the standard error: $\sqrt{\dfrac{p_o(1 - p_o)}{n}}$. Save your answer.**

4. **Divide your result from Step 2 by your result from Step 3.**

To interpret the test statistic, look up your test statistic on the standard normal distribution (see Table 8-1 in Chapter 8) and calculate the p-value (see Chapter 14 for more on p-value calculations).

For example, suppose Cavifree toothpaste claims that four out of five dentists recommend Cavifree toothpaste to their patients. In this case, the population is all dentists, and p is the proportion of all dentists who recommended Cavifree to their patients. The claim is that p is equal to "four out of five," which means that p_0 is $4 \div 5 = 0.80$. You suspect that the proportion is actually less than 0.80. Your hypotheses are H_o: $p = 0.80$ versus H_a: $p < 0.80$. Suppose that 150 out of 200 dental patients sampled received a recommendation for Cavifree.

To find the test statistic, start with $\hat{p}$ is $150 \div 200 = 0.75$. Also, $p_0 = 0.80$ and $n = 200$.

Take $0.75 - 0.80 = -0.05$.

Next, the standard error is the square root of $[(0.80 \times [1 - 0.80]) \div 200] =$ the square root of $(0.16 \div 200) =$ the square root of $0.0008 = 0.028$.

The test statistic is -0.05 divided by 0.028, which is $-0.05 \div 0.028 = -1.79$, rounded to -1.8.

This means that your sample results are 1.8 standard errors below the claimed value for the population.

How often would you expect to get results like this if H_o were true? The percentage chance of being at or beyond (in this case to the left of) -1.8, is 3.59%. (Look up -1.8 in Table 8-1 in Chapter 8 and use the corresponding percentile, because H_a is a less-than hypothesis. See Chapter 14 for more on this.) Now divide by 100 to get your p-value, which is 0.0359. Because the p-value is less than 0.05, you have enough evidence to reject H_o.

According to your sample, the claim of four out of five (80% of) dentists recommending Cavifree toothpaste is not true; the actual percentage of recommendations is less than that.

Most hypothesis tests involving proportions are done using samples that are quite large, given that they're most often based on surveys, so you rarely encounter a situation in which a very small sample is used. For information on how to calculate the p-value for the greater-than or not-equal-to alternatives, see Chapter 14.

Comparing Two (Separate) Population Averages

This test is used when the variable is numerical (for example, income, cholesterol level, or miles per gallon) and two populations or groups are being compared (for example, men versus women, athletes versus non-athletes, or cars versus SUVs). Two separate random samples need to be selected, one from each population, in order to collect the data needed for this test. The null hypothesis is that the two population means are the same; in other words, that their difference is equal to 0. The notation for the null hypothesis is $H_o: \mu_x - \mu_y = 0$, where μ_x represents the mean of the first population, and μ_y represents the mean of the second population.

The formula for the test statistic comparing two means is $\dfrac{\bar{x} - \bar{y}}{\sqrt{\dfrac{s_x^2}{n_1} + \dfrac{s_y^2}{n_2}}}$. To calculate it, do the following:

1. **Calculate the sample means ($\bar{x}$ and $\bar{y}$) and sample standard deviations (s_x and s_y) for each sample separately. Let n_1 and n_2 represent the two sample sizes (they need not be equal).**

   See Chapter 4 for these calculations.

2. **Find the difference between the two sample means, $\bar{x} - \bar{y}$.**

3. **Calculate the standard error, $\sqrt{\dfrac{s_x^2}{n_1} + \dfrac{s_y^2}{n_2}}$. Save your answer.**

4. **Divide your result from Step 2 by your result from Step 3.**

To interpret the test statistic, look up your test statistic on the standard normal distribution (see Table 8-1 in Chapter 8) and calculate the *p*-value (see Chapter 14 for more on *p*-value calculations).

For example, suppose you want to compare the absorbency of two brands of paper towels (call the brands Stats-absorbent and Sponge-o-matic). You can make this comparison by looking at the average number of ounces each brand can absorb before being saturated. H_o says the difference between the average absorbencies is 0 (non-existent), and H_a says the difference is not 0. In other words, $H_o: \mu_x - \mu_y = 0$ versus $H_a: \mu_x - \mu_y \neq 0$. Here, you have no indication of which paper towel may be more absorbent, so the not-equal-to alternative is the one to use. (See Chapter 14.)

Suppose you select a random sample of 50 paper towels from each brand and measure the absorbency of each paper towel. Suppose the average absorbency of Stats-absorbent (x) is 3 ounces, with a standard deviation of 0.9 ounces, and for Sponge-o-matic (y), the average absorbency is 3.5 ounces, with a standard deviation of 1.2 ounces.

Given these data, you have $\bar{x} = 3$, $s_x = 0.9$, $\bar{y} = 3.5$, $s_y = 1.2$, $n_1 = 50$, and $n_2 = 50$.

The difference between the sample means for (Stats-absorbent – Sponge-o-matic) is $(3 - 3.5) = -0.5$ ounces. (A negative difference simply means that the second sample mean was larger than the first.)

The standard error is $\sqrt{\dfrac{0.9^2}{50} + \dfrac{1.2^2}{50}} = \sqrt{\dfrac{0.81}{50} + \dfrac{1.44}{50}} = \sqrt{0.045} = 0.2121$.

Divide the difference, –0.5, by the standard error, 0.2121, which gives you –2.36, which rounds to –2.4. This is your test statistic.

To find the *p*-value, look up –2.4 on the standard normal distribution (Z-distribution) — see Table 8-1 in Chapter 8. The chance of being beyond, in this case to the left of, –2.4 is equal to the percentile, which is 0.82%. Because H_a is a not-equal-to alternative, you double this percentage to get $2 \times 0.82\% = 1.64\%$. Finally, change this to a probability by dividing by 100 to get a *p*-value of 0.0164. This *p*-value is less than 0.05. That means you do have enough evidence to reject H_o.

Your conclusion is that a statistically significant difference exists between the absorbency levels of these two brands of paper towels, based on your samples. And it looks like Sponge-o-matic comes out on top, because it has a higher average.

Being the savvy statistician you are, don't fall for those commercials that show one single sheet from one single roll of paper towels (that is, a sample size of 1) being more absorbent than another. And don't give credibility to those morning TV news shows that send producers on the street asking two or three people for information and making comparisons. Anecdotes are interesting, but they can't be generalized. A hypothesis test, done right, gives results that are both interesting *and* can be generalized. (See Chapter 14 for more on avoiding anecdotes.)

Most hypothesis tests comparing two separate population means are done using samples that are quite large, because they are most often based on surveys. However, if both samples do happen to be under 30 in size, you need to use the t-distribution (with degrees of freedom equal to $n_1 - 1$ or $n_2 - 1$, whichever is smaller) to look up the *p*-value. (See Table 14-2 in Chapter 14 for more on the t-distribution.)

Testing for an Average Difference (Paired Data)

This test is used when the variable is numerical (for example, income, cholesterol level, or miles per gallon), and the individuals in the sample are either paired up in some way (identical twins are often used) or the same people are used twice (for example, using a pre-test and post-test). Paired tests are typically used for studies in which they're testing to see whether a new treatment, technique, or method works better than an existing method, without having to worry about other factors about the subjects that may influence the results. See Chapter 17 for details.

For example, suppose a researcher wants to see whether teaching students to read using a computer game gives better results than teaching with a tried-and-true phonics method. She randomly selects 20 students and puts them into 10 pairs according to their reading readiness level, age, IQ, and so on. She randomly selects one student from each pair to learn to read via the computer game, and the other learns to read using the phonics method. At the end of the study, each student takes the same reading test. The data are shown in Table 15-1.

Table 15-1	Reading Scores for Computer Game versus the Phonics Method		
Student Pair #	Reading Score for Student under Computer Method	Reading Score for Student under Phonics Method	Paired Differences (Computer Score – Phonics Score)
1	85	80	+5
2	80	80	+0
3	95	88	+7
4	87	90	−3
5	78	72	+6
6	82	79	+3
7	57	50	+7
8	69	73	−4
9	73	78	−5
10	99	95	+4

The data are in pairs, but you're really interested only in the difference in reading scores (computer reading score – phonics reading score) for each pair, not the reading scores themselves. So, you take the difference between the scores for each pair, and those *paired differences* make up your new set of data to work with. If the two reading methods are the same, the average of the paired differences should be 0. If the computer method is better, the average of the paired differences should be positive (because the computer reading score should be larger than the phonics score). So you really have a hypothesis test for one population mean, where the null hypothesis is that the mean (of the paired differences) is 0, and the alternative hypothesis is that the mean (of the paired differences) is > 0.

The notation for the null hypothesis is $H_o : \mu_d = 0$, where μ_d is the mean of the paired differences. (The *d* in the subscript is just supposed to remind you that you're working with the paired differences.)

The formula for the test statistic for paired differences is $\frac{\overline{d}-0}{s/\sqrt{n}}$. To calculate it, do the following:

1. **For each pair of data, take the first value in the pair minus the second value in the pair to find the paired difference.**

   Think of the differences as your new data set.

2. **Calculate the mean, $\overline{d}$, and the standard deviation, s, of all the differences.**

   Let n represent the number of paired differences that you have.

3. **Calculate the standard error: $s/\sqrt{n}$. Save your answer.**

4. **Take $\overline{d}$ divided by the standard error from Step 3.**

   Remember that $\mu_d = 0$ if H_o is true, so it's not included in the formula here.

For the reading scores example, you can use the preceding steps to see whether the computer method is better in terms of teaching students to read.

Calculate the differences for each pair; you can see those differences in column 4 of Table 15-1. Notice that the sign on each of the differences is important; it indicates which method performed better for that particular pair.

The mean and standard deviation of the differences (column 4 of Table 15-1) must be calculated. (See Chapter 4 for calculating means and standard deviations.) The mean of the differences is found to be +2, and the standard deviation is 4.64. Note that $n = 10$ here.

The standard error is 4.64 divided by the square root of 10 (3.16). So you have $4.64 \div 3.16 = 1.47$. (Remember that here, n is the number of pairs, which is 10.)

For the last step, take the mean of the differences, +2, divided by the standard error, which is 1.47, to get +1.36, the test statistic. That means the average difference for this sample is 1.36 standard errors above 0. Is this enough to say that a difference in reading scores applies to the whole population in general?

Because n is less than 30, you look up 1.36 on the t-distribution with $10 - 1 = 9$ degrees of freedom (see Table 14-2 in Chapter 14) to calculate the p-value. The p-value in this case is greater than 0.05 because 1.36 is close to the value of 1.38 on the table, and, therefore, its p-value would be more than 0.10 (the corresponding p-value for 1.38). That's because 1.38 is in the column under the 90th percentile, and because H_a is a greater-than alternative, you take $100\% - 90\% = 10\% = 0.10$. You conclude that there isn't enough evidence to reject H_o, so the computer game can't be touted as a better reading method. (This could be due to the lack of additional evidence needed to prove the point with a smaller sample size.)

In many paired experiments, the data sets will be small due to costs and time associated with doing these kinds of studies. That means the t-distribution (see Table 14-2 in Chapter 14) is often used instead of the standard normal distribution (see Table 8-1 in Chapter 8), when figuring out the *p*-value.

Comparing Two Population Proportions

This test is used when the variable is categorical (for example, smoker/nonsmoker, political party, support/oppose an opinion, and so on) and you're interested in the proportion of individuals with a certain characteristic — for example, the proportion of smokers. In this case, two populations or groups are being compared (such as the proportion of female smokers versus male smokers). In order to conduct this test, two separate random samples need to be selected, one from each population. The null hypothesis is that the two population proportions are the same; in other words, that their difference is equal to 0. The notation for the null hypothesis is H_o: $p_1 - p_2 = 0$, where p_1 is the proportion from the first population, and p_2 is the proportion from the second population.

The formula for the test statistic comparing two proportions is

$$\frac{(\hat{p}_1 - \hat{p}_2) - 0}{\sqrt{\hat{p}(1 - \hat{p})\left(\frac{1}{n_1} + \frac{1}{n_2}\right)}}.$$ To calculate it, do the following:

1. **Calculate the sample proportions $\hat{p}_1$ and $\hat{p}_2$ for each sample. Let n_1 and n_2 represent the two sample sizes (they need not be equal).**

2. **Find the difference between the two sample proportions, $(\hat{p}_1 - \hat{p}_2)$.**

3. **Calculate the overall sample proportion, $\hat{p}$, which is the total number of individuals from both samples who have the characteristic of interest (for example, the total number of smokers, male or female, in the sample), divided by the total number of individuals from both samples $(n_1 + n_2)$.**

4. **Calculate the standard error: $\sqrt{\hat{p}(1 - \hat{p})\left(\frac{1}{n_1} + \frac{1}{n_2}\right)}$. Save your answer.**

5. **Divide your result from Step 2 by your result from Step 4.**

To interpret the test statistic, look up your test statistic on the standard normal distribution (Table 8-1 in Chapter 8) and calculate the *p*-value (see Chapter 14 for more on *p*-values).

For example, consider those drug ads that pharmaceutical companies put in magazines. The front page of an ad shows a serene picture of the sun shining, flowers blooming, people smiling — their lives changed by the drug. The company claims that its drugs can reduce allergy symptoms, help people sleep better, lower blood pressure, or fix whichever other ailment it's targeted to help. The claims may sound too good to be true, but when you turn the page to the back of the ad, you see all the fine print where the drug company justifies how it's able to make its claims. (This is typically where statistics are buried!) Somewhere in the tiny print, you'll likely find a table that shows adverse effects of the drug when compared to a control group (subjects who take a fake drug, for fair comparison to those who actually took the real drug. See Chapter 17 for more on this). For example Adderall, a drug for attention deficit hyperactivity disorder (ADHD), reported that 26 of the 374 subjects (7%) who took the drug experienced vomiting as a side effect, compared to 8 of the 210 subjects (4%) who were on a *placebo* (fake drug). Note that patients didn't know which treatment they were given. In the sample, more people on the drug experienced vomiting, but is this percentage enough to say that the entire population would experience more vomiting? You can test it to see.

In this example, you have H_o: $p_1 - p_2 = 0$ versus H_o: $p_1 - p_2 > 0$, where p_1 represents the proportion of subjects who vomited using Adderall, and p_2 represents the proportion of subjects who vomited using the placebo.

Why does H_a contain a ">" sign and not a "<" sign? H_a represents the scenario in which those taking Adderall experience more vomiting than those on placebo — that's something the FDA would want to know about. But the order of the groups is important, too. You want to set it up so the Adderall group is first, so that when you take the Adderall proportion minus the placebo proportion, you get a positive number if H_a is true. If you switch the groups, the sign would have been negative.

The next step is calculating the test statistic:

First, $\hat{p}_1 = 26 \div 374 = 0.07$ and $\hat{p}_2 = 8 \div 210 = 0.04$. The sample sizes are $n_1 = 374$ and $n_2 = 210$, respectively.

Next, take the difference between these sample proportions to get $0.07 - 0.04 = 0.03$.

The overall sample proportion, $\hat{p}$, is $(26 + 8) \div (374 + 210) = 34 \div 584 = 0.058$.

The standard error is $\sqrt{0.058 \times (1 - 0.058) \times \left(\frac{1}{374} + \frac{1}{210} \right)} = \sqrt{0.058 \times 0.942 \times 0.0074} = \sqrt{0.0004} = 0.02$. Whew!

Finally, take the difference from Step 2, 0.03, divided by 0.02 to get $0.03 \div 0.02 = 1.5$, which is the test statistic.

The p-value is the percentage chance of being at or beyond (in this case to the right of) 1.5, which is $100\% - 93.32\% = 6.68\%$, which is written as a probability as 0.0668. (See Table 8-1 of Chapter 8.) This p-value is just a shade over 0.05, so, technically, you don't have quite enough evidence to reject H_o. That means vomiting is not experienced any more by those taking this drug when compared to a placebo (although this result is one a statistician would call *marginal*).

A p-value that is very close to that magical but somewhat arbitrary cut-off of 0.05 is what statisticians call a *marginal result*. In the preceding example, the p-value of 0.0668 is generally viewed as a marginal result. It means the result is right on the borderline between accepting and rejecting H_o. That's the beauty of reporting a p-value, though; you can look at it and decide for yourself what you should conclude. The smaller the p-value, the more evidence you have against H_o, but how much evidence is enough evidence? Each person is different. If you come across a report from a study in which someone found a statistically significant result, and that result is important to you, ask for the p-value so that you can make your own decision. See Chapter 14 for more.

Most hypothesis tests comparing two separate population proportions are done using samples that are quite large, given that they're most often based on surveys, so you won't likely run into a case that uses small samples.

Part VII
Statistical Studies: The Inside Scoop

The 5th Wave By Rich Tennant

"Ted and I spent over 120 man-hours together analyzing the survey data, and here's what we discovered: Ted borrows pens and never returns them, he intentionally squeaks his chair to annoy me, and, evidently, I talk in my sleep."

In this part . . .

Many statistics that you hear and see each day are based on the results of surveys, experiments, and observational studies. Unfortunately, you can't believe everything you read or hear.

In this part, you look at what actually happens behind the scenes of these studies: how they are designed and conducted; how the data is (supposed to be) collected; and how to spot misleading results.

Chapter 16

Polls, Polls, and More Polls

Surveys seem to be all the rage in today's information explosion. Everyone wants to know how the public feels about issues, from prescription drug prices and methods of disciplining children to approval ratings of the president and reality TV. Polls and surveys are a big part of American life; they're a vehicle for quickly getting information about how you feel, what you think, and how you live your life, and are a means of quickly disseminating information about important issues. Surveys are used to highlight controversial topics, raise awareness, make political points, stress the importance of an issue, and educate or persuade the public.

Survey results can be powerful, because when many people hear that "such and such percentage of the American people do this or that," they accept these results as the truth, and then make decisions and form opinions based on that information. In fact, many surveys don't provide correct, complete, or even fair and balanced information. In this chapter, I discuss the impact of surveys and how they're used, and I take you behind the scenes of how surveys are designed and conducted so that you know what to watch for when examining survey results. I also talk about how to interpret survey results and how to spot biased and inaccurate information, so that you can determine for yourself which results to believe and which to ignore.

Recognizing the Impact of Polls

A survey is an instrument that collects data through questions and answers and is used to gather information about the opinions, behaviors, demographics, lifestyles, and other reportable characteristics of the population of interest. What's the difference between a poll and a survey? Statisticians don't make a clear distinction between the two, but I've noticed that what people call a poll is typically a short survey containing only a few questions (maybe that's how researchers get more people to respond — they call it a poll rather than a survey!). But for all intents and purposes, surveys and polls are the same thing.

You come into contact with surveys and their results on a daily basis. Surveys even have their own television program: The game show *Family Feud* is completely based on surveys and the ability of the contestants to list the top answers that people provided on a survey. Contestants on this show must correctly identify the answers provided by respondents to survey questions such as, "Name an animal you may see at the zoo" or "Name a famous person named John."

Compared to other types of studies, such as medical experiments, surveys are relatively easy to conduct and aren't as expensive to carry out. They provide quick results that can often make interesting headlines in newspapers or eye-catching stories in magazines. People connect with surveys because they feel that survey results represent the opinions of people just like themselves (even though they may never have been asked to participate in a survey). And many people enjoy seeing how other people feel, what they do, where they go, and what they care about. Looking at survey results makes people feel connected with a bigger group, somehow. That's what *pollsters* (the people who conduct surveys) bank on, and that's why they spend so much time doing surveys and polls and reporting the results of this research.

Getting to the source

Who conducts surveys these days? Pretty much anyone and everyone who has a question to ask. Some of the groups that conduct polls and report the results include:

- News organizations (for example, ABC News, CNN, Reuters)
- Political parties (those in office and those trying to get into office)
- Professional polling organizations (such as The Gallup Organization, The Harris Poll, Zogby International, and so on)
- Representatives of magazines, TV shows, and radio programs
- Professional organizations (such as the American Medical Association, which often conducts surveys of its membership)
- Special-interest groups (such as the National Rifle Association)
- Academic researchers (who conduct studies on a huge range of topics)

✔ The U.S. government (which conducts the American Community Survey, the Crime Victimization Survey, and numerous other surveys through the Census Bureau)

✔ Joe Public (who can easily conduct his own survey on the Internet)

Not everyone who conducts a poll is legitimate and trustworthy, so be sure to check the source of any survey in which you're asked to participate and for which you're given results. Groups that have a special interest in the results should either hire an independent organization to conduct (or at least to review) the survey, or they should offer copies of the survey questions to the public. Groups should also discuss in detail how the survey was designed and conducted, so that you can make an informed decision about the credibility of the results.

Surveying what's hot

The topics of many surveys are driven by current events, issues, and areas of interest; after all, timeliness and relevance to the public are two of the most attractive qualities of any survey. Here are just a few examples of some of the subjects being brought to the surface by today's surveys, along with some of the results being reported:

✔ Does celebrity activism influence the political opinions of the American public? (Over 90% of the American public says no, according to CBS News.)

✔ What percentage of Americans have dated someone online? (Only 6% of unmarried Internet users, according to CBS News.)

✔ Is pain something that lots of Americans have to deal with? (According to CBS News, three-quarters of people under 50 suffer pain often or at least some of the time.)

✔ How many people surf the Web to find health-related information? (About 98 million, according to The Harris Poll.)

✔ What's the current level of investor optimism? (According to a survey by The Gallup Organization, it should be called investor pessimism.)

✔ What was the worst car of the millennium? (The Yugo, according to listeners of the NPR radio show *Car Talk.*)

When you read the preceding survey results, do you find yourself thinking about what the results mean to you, rather than first asking yourself whether the results are valid? Some of the preceding survey results are more valid and accurate than others, and you should think about whether to believe the results first, before accepting them without question.

Ranking the worst cars of the millennium

You may be familiar with a radio show called *Car Talk* that's typically aired Saturday mornings on National Public Radio, and is hosted by Click and Clack, two brothers who offer wise and wacky advice to callers with strange car problems. The show's Web site regularly offers surveys on a wide range of car-related topics, such as, "Who has bumper stickers on their cars, and what do they say?" One of their recent surveys asked the question, "What do you think was the worst car of the millennium?" Thousands upon thousands of folks responded with their votes, but, of course, these folks don't represent all car owners. They represent only those who listen to the radio show, logged on to the Web site, and answered the survey question.

Just so you won't be left hanging (and I know you're dying to find out!) the results of the survey are shown in the following table. Before you look at what others reported, you may want to cast your own vote! (Remember, though, that these results represent only the opinions of *Car Talk* fans who took the time to get to the Web site and take the survey.) Notice that the percentages won't add up to 100% because the results in the table represent only the top ten vote-getters.

Rank	Type of Car	Percentage of Votes
1	Yugo	33.7%
2	Chevy Vega	15.8%
3	Ford Pinto	12.6%
4	AMC Gremlin	8.5%
5	Chevy Chevette	7.0%
6	Renault LeCar	4.3%
7	Dodge Aspen / Plymouth Volare	4.1%
8	Cadillac Cimarron	4.0%
9	Renault Dauphine	3.6%
10	Volkswagen (VW) Bus	2.7%

Impacting lives

Whereas some surveys are fun to look at and think about, other surveys can have a direct impact on your life or your workplace. These life-decision surveys need to be closely scrutinized before action is taken or important decisions are made. Surveys at this level can cause politicians to change or create new laws, motivate researchers to work on the latest problems, encourage manufacturers to invent new products or change business policies and practices, and influence people's behavior and ways of thinking. The following are some examples of recent survey results that can impact you:

✓ **Teens drive under the influence:** A recent Reuters survey of 1,119 teenagers in Ontario, Canada, from grades 7 through 13 found that, at some point during the previous year, 15% of them had driven a car after consuming at least two drinks.

✓ **Children's health care suffers:** A survey of 400 pediatricians by the Children's National Medical Center in Washington, D.C., reported that pediatricians spend, on average, only 8 to 12 minutes with each patient.

✓ **Crimes go unreported:** According to the U.S. Bureau of Justice 2001 Crime Victimization Survey, only 49.4% of violent crimes were reported to police. The reasons victims gave for not reporting crimes to the police are listed in Table 16-1.

Table 16-1   **Reasons Victims Didn't Report Violent Crimes to the Police**

Reason For Not Reporting	Percentage
Considered it to be a personal matter	19.2
The offender was not successful/didn't complete the crime	15.9
Reported the crime to another official	14.7
Didn't consider the crime to be important enough	5.5
Didn't think police would want to be bothered	5.3
Lack of proof	5.0
Fear of reprisal	4.6
Too inconvenient/time consuming to report it	3.9
Thought police would be biased/ineffective	2.7
Property stolen had no ID number	0.5
Not aware that a crime occurred until later	0.4
Other reasons	22.3

The most frequently given reason for not reporting a violent crime to the police was that the victim considered it to be a personal matter (19.2%). Note that almost 12% of the reasons relate to perception of the reporting process itself (for example, that it would take too much time or that the police would be bothered, biased, or ineffective).

✔ **Breast cancer treatment challenged:** Which treatment should a woman with breast cancer choose, lumpectomy (where the tumor is removed but most of the breast is not) or mastectomy (where the entire breast is removed)? The most popular belief has been that a lumpectomy is the best option for most women. However, in a recent survey asking surgeons what they would opt for if they themselves had early stage breast cancer, 50% would prefer a mastectomy.

✔ **Cellphones and driving create hazard:** A recent *Consumer Reports* survey of new car buyers and leasers looked at the driving-related issues that concerned these consumers the most. The survey found that the driving-related issue that concerned the largest percentage of respondents (53%) was drivers who were distracted by the use of cellphones. This beat out high gas prices (of concern to 50% of the respondents) and road rage (which was a concern for 38% of the respondents).

✔ **Cyber crime takes a toll on business:** The Computer Security Institute (CSI) conducted a recent survey of U.S. corporations to assess the extent to which cyber crime has affected their businesses. Ninety percent of respondents detected computer security breaches within the last year, and 80% of those acknowledged financial losses due to these breaches. Seventy-eight percent of respondents reported employee abuse of Internet-access privileges (for example, viewing pornography, downloading pirated software, and abusing e-mail privileges).

✔ **Sexual harassment presents workplace problem:** A recent survey of the workplace conducted by the Equal Employment Opportunity Commission found that between 40% and 70% of women and 10% to 20% of men reported that they had been victims of sexual harassment in the workplace. In recent years, the number of complaints filed by men has tripled.

The preceding examples address some very important issues, but you have to decide whether you can believe in or act on those results. You have to be able to sort out what's credible and reliable from what isn't. Rule number one, don't automatically believe everything you read!

Behind the Scenes: The Ins and Outs of Surveys

Surveys and their results are a part of your daily life, and you use these results to make decisions that affect your life. (Some decisions may even be life changing.) Looking at surveys with a critical eye is important. Before taking action or making decisions based on survey results, you must determine whether those results are credible, reliable, and believable. A good way to begin developing these detective skills is to go behind the scenes and see how surveys are designed, developed, implemented, and analyzed.

The survey process can be broken down into a series of ten steps:

1. **State the purpose of the survey.**
2. **Define the target population.**
3. **Choose the type of survey.**
4. **Design the questions.**
5. **Consider the timing of the survey.**
6. **Select the sample.**
7. **Collect the data.**
8. **Follow up, follow up, and follow up.**
9. **Organize and analyze the data.**
10. **Draw conclusions.**

Each step presents its own set of special issues and challenges, but each step is critical in terms of producing survey results that are fair and accurate. This sequence of steps helps you design, plan, and implement a survey, but it can also be used to critique someone else's survey, if those results are important to you.

Planning and designing a survey

The purpose of a survey is to answer questions about a target population. The *target population* is the entire group of individuals that you're interested in drawing conclusions about. In most situations, surveying the entire target population is impossible because researchers would have to spend too much time or money to do so. (When you do a survey of the entire target population, that's called a *census*.) Usually, the best you can do is to select a sample of individuals from the target population, survey those individuals, then draw conclusions about the target population based on the data from that sample.

Sounds easy, right? Wrong. Many potential problems arise after you realize that you can't survey everyone in the entire target population. Unfortunately many surveys are conducted without taking the time needed to think through these issues, and this results in errors, misleading results, and wrong conclusions.

Stating the purpose of the survey

This sounds like it should just be common sense, but in reality, many surveys have been designed and carried out that never met their purpose, or that met only some of the objectives, but not all of them. Getting lost in the questions and forgetting what you're really trying to find out is easy to do. In stating the purpose of a survey, be as specific as possible. Think about the types of conclusions you would want to make if you were to write a report, and let that help you determine your goals for the survey.

The more specific you can be about the purpose of the survey, the more easily you can design questions that meet your objectives and the better off you'll be when you need to write your report.

Defining the target population

Suppose, for example, that you want to conduct a survey to determine the extent to which people engage in personal e-mail usage in the workplace. You may think that the target population is e-mail users in the workplace. However, you want to determine the *extent* to which e-mail is used in the workplace, so you can't just ask e-mail users, or your results would be biased against those who don't use e-mail in the workplace. But should you also include those who don't even have access to a computer during their workday? (See how fast surveys can get tricky?)

The target population that probably makes the most sense here is all of the people who use Internet-connected computers in the workplace. Everyone in this group at least has access to e-mail, though only some of those with access to e-mail in the workplace actually use it, and of those who use it, only some use it for personal e-mail. (And that's what you want to find out—how much they do use e-mail for that purpose.)

You need to be clear in your definition of the target population. Your definition is what helps you select the proper sample, and it also guides you in your conclusions, so that you don't over-generalize your results. If the researcher didn't clearly define the target population, this can be a sign of other problems with the survey.

Choosing the type of survey

The next step in designing your survey is to choose what type of survey is most appropriate for the situation at hand. Surveys can be done over the phone, through the mail, with door-to-door interviews, or over the Internet. However, not every type of survey is appropriate for every situation. For example, suppose you want to determine some of the factors that relate to illiteracy in the United States. You wouldn't want to send a survey through the mail, because people who can't read won't be able to take the survey. In that case, a telephone interview is more appropriate.

Choose the type of survey that's most appropriate for the target population, in terms of getting the most truthful and informative data possible. When examining the results of a survey, be sure to look at whether the type of survey used is most appropriate for the situation.

Designing the questions

After the purpose of the survey has been clearly outlined and you've chosen the type of survey you're going to use, the next step is to design the questions. The way that the questions are asked can make a huge difference in the quality of the data that will be collected. One of the single most common sources of bias in surveys is the wording of the questions. *Leading questions* (questions that are designed to favor a certain response over another) can greatly affect how people answer the questions, and these responses may not accurately reflect how the respondents truly feel about an issue. For example, here are two ways that you could word a survey question about a proposed school bond issue (both of which are leading questions):

> *Don't you agree that a tiny percentage increase in sales tax is a worthwhile investment in improving the quality of the education of our children?*

> *Don't you think we should stop increasing the burden on the taxpayers and stop asking for yet another sales tax hike to fund the wasteful school system?*

From the wording of each of these leading questions, you can easily see how the pollster wants you to respond. Research shows that the wording of the question does affect the outcome of surveys. The best way to word a question is in a neutral way. For this example, the question should be worded like this:

> *The school district is proposing a 0.01% increase in sales tax to provide funds for a new high school to be built in the district. What's your opinion on the proposed sales tax? (Possible responses: strongly in favor, in favor, neutral, against, strongly against.)*

In a good survey, the questions are always worded in a neutral way in order to avoid bias. The best way to assess the neutrality of a question is to ask yourself whether you can tell how the person wants you to respond by reading the question. If the answer is yes, that question is a leading question and can give misleading results.

If the results of a survey are important to you, ask the researcher for a copy of the questions used on the survey, so you can assess the quality of the questions.

Timing the survey

In a survey, as in life, timing is everything. Current events shape people's opinions all the time, and whereas some pollsters try to determine how people feel about those events, others take advantage of events, especially negative ones, and use them as political platforms or as fodder for headlines and controversy. The timing of any survey can also cause bias, regardless of the subject matter. For example, suppose your target population for a survey is people who work full time. If you conduct a telephone survey to get office workers' opinions on personal e-mail use at work, and you call them at home between the hours of 9 a.m. and 5 p.m., you're going to have bias in your results, because those are the hours when the majority of office workers are at work.

Check out when a survey was conducted (time and date) and see whether you can determine any relevant events that occurred at that time that may have influenced the results. Also verify that the survey was conducted during a time of the day that's most convenient for the target population to respond.

Selecting the sample

After the survey has been designed, the next step is to select the people who will participate in the survey. Because typically you don't have time or money to conduct a census (a survey of the entire target population) you need to select a subset of the population, called a *sample*. How this sample is selected can make all the difference in terms of the accuracy and the quality of the results.

Three criteria are important in selecting a good sample:

✔ **A good sample represents the target population.** To represent the target population, the sample must be selected from the target population, the whole target population, and nothing but the target population. Suppose you want to find out how many hours of TV Americans watch in a day, on average. Asking students in a dorm at a local university to record their TV viewing habits isn't going to cut it. Students represent only a portion of the target population. Asking people to call in their opinions on a radio show is not going to give you a sample that represents the target population; the results will represent only the people who were listening to the show, were able to call at the appropriate time, and felt strongly enough about the issue to make the effort to call in. Likewise, a Web survey will represent only the people who have access to the Internet and who logged on to the site where the survey was posted.

Unfortunately, many people who conduct surveys don't take the time or spend the money to select a representative sample of people to participate in the study, and this leads to biased survey results. When presented with survey results, find out how the sample was selected before examining the results of the survey.

✔ **A good sample is selected randomly.** A *random* sample is one in which every member of the target population has an equal chance of being selected. The easiest example to visualize here is that of a hat (or bucket) containing individual slips of paper, each with the name of a person written on it; if the slips are thoroughly mixed before each slip of paper is drawn out, the result will be a random sample of the target population (in this case, the population of people whose names are in the hat). A random sample eliminates bias in the sampling process.

Reputable polling organizations, such as The Gallup Organization, use a random digit dialing procedure to telephone the members of their sample. Of course, this excludes people without telephones, but because most American households today have at least one telephone, the bias involved in excluding people without telephones is relatively small.

Beware of surveys that have a large sample size but where that sample is not randomly selected. Internet surveys are the biggest culprit. Someone can say that 50,000 people logged on to a Web site to answer a survey, and that means that the Webmaster of this site has a lot of information. But that information is biased, because it doesn't represent the opinions of anyone except those who knew about the survey, chose to participate in it, and had access to the Internet. In a case like this, less would have been more: This survey designer should have sampled fewer people but done so randomly.

✔ **A good sample is large enough for the results to be accurate.** If you have a large sample size, and if the sample is representative of the target population and is selected at random, you can count on that information being pretty accurate. How accurate depends on the sample size, but the bigger the sample size, the more accurate the information will be (as long as that information is good information). The accuracy of most survey questions is measured in terms of a percentage. This percentage is called the *margin of error,* and it represents how much the researcher expects the results to vary if he or she were to repeat the survey many times using different samples of the same size. Read more about this in Chapter 10.

A quick and dirty formula to estimate the accuracy of a survey is to take 1 divided by the square root of the sample size. For example, a survey of 1,000 (randomly selected) people is accurate to within $\dfrac{1}{\sqrt{1,000}} = 0.032$ or 3.2 percentage points. (Note that in cases where not everyone responded, you should replace the sample size with the number of respondents. See the "Following up, following up, and following up" section later in this chapter.)

Carrying out a survey

The survey has been designed, and the participants have been selected. Now you have to go about the process of carrying out the survey, which is another important step, one where lots of mistakes and bias can occur.

Collecting the data

During the survey itself, the participants can have problems understanding the questions, they may give answers that aren't among the choices (in the case of a multiple choice question), or they may decide to give answers that are inaccurate or blatantly false. (As an example of this third type of error, where respondents provide false information, think about the difficulties involved in getting people to tell the truth about whether they've cheated on their income-tax forms.) This third type of error is called *response bias* — the respondent gives a biased answer.

Some of the potential problems with the data-collection process can be minimized or avoided with careful training of the personnel who carry out the survey. With proper training, any issues that arise during the survey are resolved in a consistent and clear way, and no errors are made in recording the data. Problems with confusing questions or incomplete choices for answers can be resolved by conducting a pilot study on a few participants prior to the actual survey, and then, based on their feedback, fixing any problems with the questions. Personnel can also be trained to create an environment in which each respondent feels safe enough to tell the truth; ensuring that privacy will be protected also helps encourage more people to respond.

Following up, following up, and following up

Anyone who has ever thrown away a survey or refused to "answer a few questions" over the phone knows that getting people to participate in a survey isn't easy. If the researcher wants to minimize bias, the best way to handle this is to get as many folks to respond as possible by following up, one, two, or even three times. Offer dollar bills, coupons, self-addressed stamped return envelopes, chances to win prizes, and so on. Every little bit helps.

What has ever motivated you to fill out a survey? If the incentive provided by the researcher didn't persuade you (or that feeling of guilt for just taking those two shiny quarters and dumping the survey in the trash didn't get to you) maybe the subject matter peaked your interest. This is where bias comes in. If only those folks who feel very strongly respond to a survey, that means that only their opinions will count, because the other people who didn't really care about the issue didn't respond, and their "I don't care" vote didn't get counted. Or maybe they did care, but they just didn't take the time to tell anyone. Either way, their vote doesn't count.

For example, suppose 1,000 people are given a survey about whether the park rules should be changed to allow dogs on leashes. Who would respond? Most likely, the respondents would be those who strongly agree or disagree with the proposed rules. Suppose 100 people from each of the two sides of the issue were the only respondents. That would mean that 800 opinions were not counted. Suppose none of those 800 people really cared about the issue either way. If you could count their opinions, the results would be $800 \div 1,000 = 80\%$ "no opinion," $100 \div 1,000 = 10\%$ in favor of the new rules and $100 \div 1,000 = 10\%$ against the new rules. But without the votes of the 800 non-respondents, the researchers would report, "Of the people who responded, 50% were in favor of the new rules and 50% were against them." This gives the impression of a very different (and a very biased) result from the one you would've gotten if all 1,000 people had responded.

Lying: What do they know?

A study published in the *Journal of Applied Social Psychology* concluded that when lying to someone is in the best interest of the person hearing the lie, lying becomes more socially acceptable, and when lying to someone is in the best interest of the liar himself/herself, the lying becomes less socially acceptable. This sounds interesting and seems to make sense, but can this be true of everyone? The way the results are stated, this appears to be the case. However, in looking at the people who actually participated in the survey leading to these results, you begin to get the feeling that the conclusions may be a bit ambitious, to say the least.

The authors started out with 1,105 women who were selected to participate in the survey. Of these, 659 refused to cooperate, most of them saying they didn't have time. Another 233 were determined by the researchers to be either "too young" or "too old," and 33 were deemed unsuitable because the researchers sited a language barrier. In the end, 180 women were questioned. The average age of the final group of participants was 34.8 years.

Wow, where do I start with this one? The original sample size of 1,105 seems large enough, but were they selected randomly? Notice that the entire sample consisted of women, which is interesting, because the conclusions don't say

that lies are more or less acceptable according to women in these situations. Next, 659 of those selected to participate refused to cooperate (causing bias). This is a large percentage (60%), but given the subject matter, you shouldn't be surprised. The researchers could have minimized the problem by guaranteeing that the responses would be anonymous, for example. No information is given regarding whether any follow-up was done (which probably means that it wasn't).

Throwing out 233 people because they were "too young" or "too old" is just plain wrong, unless your target population is limited to a certain age group. If that had been the case, the conclusions should have been made about that age group only. Finally, the last straw: throwing out 33 people who were "unsuitable" (in the researchers' own terms) because of a language barrier. I'd say, "Bring in an interpreter," because the conclusions were not limited to only those who speak English. You can't have the survey participants represent only a tiny microcosm of society (young women who speak English), and then turn around and make conclusions about all of society, based only on the data from this tiny microcosm. Starting out with a sample size of 1,105 and ending up with only 180 women is just plain bad statistics.

The *response rate* of a survey is a ratio found by taking the number of respondents divided by the number of people who were originally asked to participate. Statisticians feel that a good response rate is anything over 70%. However, many response rates fall far short of that, unless the survey is done by a reputable organization, such as The Gallup Organization. Look for the response rate when examining survey results. If the response rate is too low (much less than 70%) the results may be biased and should be ignored. Don't be fooled by a survey that claims to have a large number of respondents but actually has a low response rate; in this case, many people may have responded, but many more were asked and didn't respond.

Note that many statistical formulas (including the formulas in this book) assume that your sample size is equal to the number of respondents, because statisticians want you to know how important it is to follow up with people and not end up with biased data due to non-response. However, in reality, statisticians know that you can't always get everyone to respond, no matter how hard you try. So, which number do you put in for *n* in all the formulas: the intended sample size (the number of people contacted) or the actual sample size (the number of people who responded)? Use the number of people who responded. Note, however, that for any survey with a low response rate, the results shouldn't be reported, because they very well could be biased. That's how important following up really is. (Do other people heed this warning when they report their results to you? Not often enough.)

Regarding the quality of results, selecting a smaller initial sample size and following them up more aggressively is a much better approach than selecting a larger group of potential respondents and having a low response rate.

Interpreting results; detecting problems

The purpose of a survey is to gain information about your target population; this information can include opinions, demographic information, or lifestyles and behaviors. If the survey has been designed and conducted in a fair and accurate manner with the goals of the survey in mind, the data should provide good information as to what's happening with the target population (within the stated margin of error). The next steps are to organize the data to get a clear picture of what's happening; analyze the data to look for links, differences, or other relationships of interest; and then to draw conclusions based on the results.

Organizing and analyzing

After a survey has been completed, the next step is to organize and analyze the data (in other words, crunch some numbers and make some graphs). Many different types of data displays and summary statistics can be created and calculated from survey data, depending on the type of information that was collected. (Numerical data, such as income, have different characteristics and are usually presented differently than categorical data, such as gender.) For more information on how data can be organized and summarized, see Chapters 4 and 5 (respectively). Depending on the research question, different types of analyses can be performed on the data, including coming up with population estimates, testing a hypothesis about the population, or looking for relationships, to name a few. See Chapters 13, 15, and 18 for more on each of these analyses, respectively.

Watch for misleading graphs and statistics. Not all survey data are organized and analyzed fairly and correctly. See Chapter 2 for more about how statistics can go wrong.

Anonymity versus confidentiality

If you were to conduct a survey to determine the extent of personal e-mail usage at work, the response rate would probably be an issue, because many people are reluctant to discuss their use of personal e-mail in the workplace, or at least to do so truthfully. You could try to encourage people to respond by letting them know that their privacy would be protected during and after the survey.

When you report the results of a survey, you generally don't tie the information collected to the names of the respondents, because doing so would violate the privacy of the respondents. You've probably heard the terms "anonymous" and "confidential" before, but what you may not realize is that these two words are completely different in terms of privacy issues. Keeping results *confidential* means that I could tie your information to your name in my report, but I promise that I won't do that. Keeping results *anonymous* means that I have no way of tying your information to your name in my report, even if I wanted to.

If you're asked to participate in a survey, be sure you're clear about what the researchers plan to do with your responses and whether or not your name can be tied to the survey. (Good surveys always make this issue very clear for you.) Then make a decision as to whether you still wish to participate.

Drawing conclusions

The conclusions are the best part of any survey; this is why the researchers do all of the work in the first place. If the survey was designed and carried out properly, the sample was selected carefully — and the data were organized and summarized correctly — the results will fairly and accurately represent the reality of the target population. But, of course, not all surveys are done right. But even if a survey is done correctly, researchers can misinterpret or over-interpret results so that they say more than they really should. You know the saying, "Seeing is believing"? Some researchers are guilty of the converse, which is, "Believing is seeing." In other words, they claim to see what they want to believe about the results. All the more reason for you to know where the line is drawn between reasonable conclusions and misleading results, and to realize when others have crossed that line.

Here are some of the most common errors made in drawing conclusions from surveys:

- Making projections to a larger population than the study actually represents
- Claiming a difference exists between two groups when a difference isn't really there
- Saying that "these results aren't scientific, but. . . ," and then going on to present the results as if they are scientific

Getting too excited?

In 1998, a press release put out by the search engine Excite stated that they were named the best-liked Web site in a *USA Today* study conducted by Intelliquest. The survey was based on 300 Web users selected from a group of 30,000 technology panelists who worked for Intelliquest. (Note that this is *not* a random sample of Web users!) The conclusions stated that Excite won the overall consumer experience category, making it the best-liked site on the Web, beating out Yahoo! and the other competitors.

Excite claimed that it was better than Yahoo! based on this survey. The actual results, however, tell a different story. The average overall quality score, on a scale of 0 to 100%, was 89% for Excite and 87% for Yahoo!. Excite's score is admittedly good, and it is slightly higher than the score obtained for Yahoo!; however, the difference between the results for the two companies are actually well within the margin of error for this survey, which is plus or minus 3.5%. In other words, Excite and Yahoo! were in a statistical tie for first place. So saying which company actually came in first isn't possible in this case. (See Chapters 9 and 10 for more on sample variation and the margin or error.)

To avoid common errors made when drawing conclusions, do the following:

1. **Check whether the sample was selected properly and that the conclusions don't go beyond the population presented by that sample.**

2. **Look for disclaimers about surveys *before* reading the results, if you can.**

   That way, you'll be less likely to be influenced by the results you're reading, if, in fact, the results aren't based on a scientific survey. Now that you know what a *scientific survey* (the media's term for an accurate and unbiased survey) actually involves, you can use those criteria to judge for yourself whether the survey results are credible.

3. **Be on the lookout for statistically incorrect conclusions.**

   If someone reports a difference between two groups in terms of survey results, be sure that the difference is larger than the reported margin of error. If the difference is within the margin of error, you should expect the sample results to vary by that much just by chance, and the so-called "difference" can't really be generalized to the entire population. (See Chapter 14 for more on this.)

4. **Tune out anyone who says, "These results aren't scientific, but. . . ."**

 Here's the bottom line about surveys. Know the limitations of any survey and be wary of any information coming from surveys in which those limitations aren't respected. A bad survey is cheap and easy to do, but you get what you pay for. Before looking at the results of any survey, investigate how it was designed and conducted, so that you can judge the quality of the results.

Chapter 17

Experiments: Medical Breakthroughs or Misleading Results?

In This Chapter

▶ Considering the limitations of observational studies

▶ Finding out how experiments work

▶ Watching for misleading results

Medical breakthroughs seem to come and go quickly in today's age of information. One day, you hear about a promising new treatment for a disease, only to find out later that the drug didn't live up to expectations in the last stage of testing. Pharmaceutical companies bombard TV viewers with commercials for pills, sending millions of people to their doctors clamoring for the latest and greatest cures for their ills, sometimes without even knowing what the drugs are for. Anyone can search the Internet for details about any type of ailment, disease, or symptom and come up with tons of information and advice. But how much can you really believe? And how do you decide which options are best for you if you get sick, need surgery, or have an emergency?

In this chapter, you get behind the scenes of experiments, the driving force of medical studies and other investigations in which comparisons are made — comparisons that test, for example, which building materials are best, which soft drink teens prefer, which SUV is safest in a crash, and so on. You find out the difference between experiments and observational studies and discover what experiments can do for you, how they're supposed to be done, how they can go wrong, and how you can spot misleading results. With so many news headlines, sound bites, and pieces of "expert advice" coming at you from all directions, you need to use all of your critical thinking skills to evaluate the sometimes-conflicting information you're presented with on a regular basis.

Determining What Sets Experiments Apart

Although many different types of studies exist, you can boil them all down to basically two different types: experiments and observational studies. This section examines what, exactly, makes experiments different from other studies.

An *observational study* is just what it sounds like: a study in which the researcher merely observes the subjects and records the information. No intervention takes place, no changes are introduced, and no restrictions or controls are imposed. An *experiment* is a study that doesn't simply observe subjects in their natural state, but deliberately applies treatments to them in a controlled situation and records the outcomes.

Examining experiments

The basic goal of an experiment is to find out whether a particular treatment causes a change in the response. (The operative word here is "cause.") The way an experiment does this is by creating a very controlled environment — so controlled that the researcher can pinpoint whether a certain factor or combination of factors causes a change in the response variable, and if so, the extent to which that factor (or that combination of factors) influences the response.

For example, in order to gain government approval for a proposed drug, pharmaceutical researchers set up experiments to determine whether that drug helps lower blood pressure, what dosage level is most appropriate for each different population of patients, what side effects (if any) occur, and to what extent those side effects occur in each population.

Observing observational studies

In certain situations, observational studies are the optimal way to go. The most common observational studies are polls and surveys (see Chapter 16). When the goal is simply to find out what people think and to collect some demographic information (such as gender, age, income, and so on), surveys and polls can't be beat, as long as they're designed and conducted correctly.

In other situations, especially those looking for cause-and-effect relationships (discussed in detail in Chapter 18), observational studies aren't appropriate. For example, suppose you took a couple of Vitamin C pills last week; is that what helped you avoid getting that cold that's going around the office? Maybe the extra sleep you got recently or the extra hand-washing you've been doing helped you ward off the cold. Or maybe you just got lucky this time. With so many variables in the mix, how can you tell which one had an influence on the outcome of your not getting a cold?

When looking at the results of any study, first determine what the purpose of the study was and whether the type of study fits the purpose. For example, if an observational study was done instead of an experiment in order to establish a cause-and-effect relationship (see Chapter 18), any conclusions that are drawn should be carefully scrutinized.

Respecting ethical issues

The trouble with experiments is that some experimental designs are not always ethical. That's why so much evidence was needed to show that smoking causes lung cancer, and why the tobacco companies only recently had to pay huge penalties to victims. You can't force research subjects to smoke in order to see what happens to them. You can only look at people who have lung cancer and work backward to see what *factors* (variables being studied) may have caused the disease. But because you can't control for the various factors you're interested in — or for any other variables, for that matter — singling out any one particular cause becomes difficult with observational studies.

Although the causes of cancer and other diseases can't be determined ethically by conducting experiments on humans, treatments for cancer can be (and are) tested using experiments. Medical studies that involve experiments are called *clinical trials*. Check out www.clinicaltrials.gov for more information.

Surveys, polls, and other observational studies are fine if you want to know people's opinions, examine their lifestyles without intervention, or examine some demographic variables. If you want to try to determine the cause of a certain outcome or behavior (that is, a reason why something happened), an experiment is a much better way to go. If an experiment isn't possible (because it's unethical, too expensive, or otherwise unfeasible), a large body of observational studies examining many different factors and coming up with similar conclusions is the next best thing. (See Chapter 18 for more about cause-and-effect relationships.)

Designing a Good Experiment

How an experiment is designed can mean the difference between good results and garbage. Because most researchers are going to write the most glowing press releases that they can about their experiments, you have to be able to sort through the hype in order to determine whether to believe the results you're being told.

To decide whether an experiment is credible, check to see whether it follows these steps for a good experiment:

1. **Includes a large enough sample size so that the results are accurate.**

2. **Chooses subjects that most accurately represent the target population.**

3. **Assigns subjects randomly to the treatment group(s) and the control group.**

4. **Controls for possible confounding variables.**

5. **Double-blinds the study to avoid bias.**

6. **Collects good data.**

7. **Contains the proper data analysis.**

8. **Doesn't make conclusions that go beyond the scope and limitations of the study.**

In the following sections, each of these criteria is explained in more detail and illustrated using various examples.

Selecting the sample size

The size of a sample greatly affects the accuracy of the results. The larger the sample size, the more accurate the results and the more powerful the statistical tests (in terms of being able to detect real results when they exist). See Chapters 10 and 14 for more details.

Understanding that small samples don't make for big conclusions

You may be surprised at the number of research headlines that have been made that were based on very small samples. This is an issue of great concern to statisticians, who know that in order to detect most differences between groups you need sample sizes that are large (at least larger than 30; see Chapter 10). When sample sizes are small and big conclusions have been made, either the researchers didn't use the right hypothesis test to analyze their data (they often should be using the t-distribution rather than the Z-distribution; see Chapter 14) or the difference was so large that a small sample size was all that was needed to detect that difference. The latter usually isn't the case, however.

 Be wary of research conclusions that find significant results based on small sample sizes (especially samples much smaller than 30). If the results are important to you, ask for a copy of the research report and look to see what type of analysis was done on the data. Also look at the sample of subjects to see whether this sample truly represents the population about which the researchers are drawing conclusions.

Checking your definition of "sample size"

When asking questions about sample sizes, be specific about what you mean by sample size. For example, you can ask how many subjects were selected to participate and also ask for the number who actually completed the experiment; these two numbers can be very different. Make sure the researchers can explain any situations in which the research subjects decided to drop out or were unable (for some reason) to finish the experiment.

For example, an article in the *New York Times* entitled "Marijuana Is Called an Effective Relief in Cancer Therapy" says in the opening paragraph that marijuana is "far more effective" than any other drug in relieving the side effects of chemotherapy. When you get into the details, you find out that the results are based on only 29 patients (15 on the treatment, 14 on a placebo). To add to the confusion, you find out that only 12 of the 15 patients in the treatment group actually completed the study; so what happened to the other three subjects?

Sometimes, researchers draw their conclusions based on only those subjects who completed the study. This can be misleading, because the data don't include information about those who dropped out (and why). This can lead to biased data. For a discussion of the sample size you need in order to achieve a certain level of accuracy, see Chapter 12.

Accuracy isn't the only issue in terms of having "good" data. You still need to worry about eliminating bias by selecting a random sample (see Chapter 3 for more on how random samples are taken).

Choosing the subjects

The first step in designing an experiment is selecting the sample of participants, called the research *subjects*. Although researchers would like for their subjects to be selected randomly from their respective populations, in most cases, this just isn't feasible. For example, suppose a group of eye researchers wants to test out a new laser surgery on nearsighted people. They need a random sample of subjects, so they randomly select various eye doctors from across the country and randomly select nearsighted patients from these doctors' files. They call up each person selected and say, "We're experimenting with a new laser surgery treatment for nearsightedness, and you've been selected at random to participate in our study. When can you come in for the surgery?"

Something tells me that this approach wouldn't go over very well with many people receiving the call (although some would probably jump at the chance, especially if they didn't have to pay for the procedure). The point is, getting a truly random sample of people to participate in an experiment is generally more difficult than getting a random sample of folks to participate in a survey.

Volunteering can have side effects

In order to find subjects for their experiments, researchers often advertise for volunteers and offer them incentives such as money, free treatments, or follow-up care for their participation. Medical research on humans is complicated and difficult, but it's necessary in order to really know whether a treatment works, how well it works, what the dosage should be, and what the side effects are. In order to prescribe the right treatments in the right amounts in real-life situations, doctors and patients depend on these studies being representative of the general population. In order to recruit such representative subjects, researchers have to do a broad advertisement campaign and select enough participants with enough different characteristics to represent a cross-section of the populations of folks who will be prescribed these treatments in the future.

The U.S. National Institutes of Health has a Web site (www.clinicaltrials.gov) providing current information about clinical research studies.

Randomly assigning subjects to groups

After the sample has been selected, the researchers divide the research subjects into different groups. The subjects are assigned to one or more *treatment groups,* which receive various dosage levels of the drug or treatment being studied, and a *control group,* which receives either no treatment or a fake treatment.

Realizing the importance of random assignment

Suppose a researcher wants to determine the effects of exercise on heart rate. The subjects in his treatment group run five miles and have their heart rates measured before and after the run. The subjects in his control group will sit on the couch the whole time and watch reruns of *The Simpsons.* Which group would you rather be in? Some of the health nuts out there would no doubt volunteer for the treatment group. If you're not crazy about the idea of running five miles, you may opt for the easy way out and volunteer to be a couch potato. (Or maybe you hate *The Simpsons* so much that you'd even run five miles to avoid watching an episode.) What impact would this selective volunteering have on the results of the study? If only the health nuts (who probably already have excellent heart rates) volunteer to be in the treatment group, the researcher will be looking only at the effect of the treatment (running five miles) on very healthy and active people. He won't see the effect that running five miles has on the heart rates of couch potatoes. This non-random assignment of subjects to the treatment and control group could have a huge impact on the conclusions he draws from this study.

In order to avoid major bias in the results of an experiment, subjects must be randomly assigned to treatments and not allowed to choose which group they will be in. Keep this in mind when you evaluate the results of an experiment.

Controlling for the placebo effect

A fake treatment takes into account what researchers call the placebo effect. The *placebo effect* is a response that people have (or think they're having) just because they believe they're getting some sort of "treatment" (even if that treatment is a fake treatment, such as sugar pills). See Chapter 3 for more on the placebo effect.

When you see an ad for a drug in a magazine, look for the fine print. Embedded there, you'll see a table that lists the side effects reported by a group who took the drug, compared to the side effects reported by the control group. If the control group is on a placebo, you may expect them not to report any side effects, but you would be wrong. Placebo groups often report side effects in percentages that seem quite high; this is because their minds are playing tricks on them and they're experiencing the placebo effect. If you want to be fair about examining the reported side effects of a treatment, you have to also take into account the side effects that the control group reports — side effects that are due to the placebo effect only.

In some situations, such as when the subjects have very serious diseases, offering a fake treatment as an option may be unethical. In 1997, the U.S. government was harshly criticized for financing an HIV study that examined dosage levels of AZT, a drug known at that time to cut the risk of HIV transmission from pregnant mothers to their babies by two-thirds. This particular study, in which 12,000 pregnant women with HIV in Africa, Thailand, and the Dominican Republic participated, had a deadly design. Researchers gave half of the women various dosages of AZT, but the other half of the women received sugar pills. Of course, had the U.S. government realized that a placebo was being given to half of the subjects, they wouldn't have supported the HIV study.

Here's the way these situations are supposed to be handled. When ethical reasons bar the use of fake treatments, the new treatment is compared to an existing or standard treatment that is known to be an effective treatment. After researchers have enough data to see that one of the treatments is working better than the other, they will generally stop the experiment and put everyone on the better treatment, again for ethical reasons.

When examining the results of an experiment, make sure the researchers compared a treatment group to a control group, in order to make sure the experiences of the treatment group go beyond what the control group experienced. The control group may receive either a fake treatment or a standard treatment, depending on the situation.

Controlling for confounding variables

Suppose you're participating in a research study that looks at factors influencing whether or not you catch a cold. If a researcher records only whether you got a cold after a certain period of time and asks questions about your behavior (how many times per day you washed your hands, how many hours of sleep you get each night, and so on), the researcher is conducting an observational study. The problem with this type of observational study is that without controlling for other factors that may have had an influence and without regulating which action you were taking when, the researcher won't be able to single out exactly which of your actions (if any) actually impacted the outcome.

The biggest limitation of observational studies is that they can't really show true cause-and-effect relationships, due to what statisticians call confounding variables. A *confounding variable* is a variable or factor that was not controlled for in the study, but can have an influence on the results.

For example, one news headline boasted, "Study links older mothers, long life." The opening paragraph said that women who have a first baby after age 40 have a much better chance of living to be 100, compared to women who have a first baby at an earlier age. When you get into the details of the study (done in 1996) you find out, first of all, that it was based on 78 women in suburban Boston who had lived to be at least 100, compared to 54 women who were born at the same time (1896), but died in 1969 (the earliest year the researchers could get computerized death records). This so-called "control group" lived to be exactly 73, no more and no less. Of the women who lived to be at least 100 years of age, 19% had given birth after age 40, whereas only 5.5% of the women who died at age 73 had given birth after age 40.

I have a real problem with these conclusions. What about the fact that the "control group" was based only on mothers who died in 1969 at age 73? What about all of the other mothers who died before age 73, or who died between the ages of 73 and 100? Maybe the control group (being so limited in scope) included women who had some sort of connection; maybe that connection caused many of them to die in the same year, and maybe that connection is further linked to why more of them had babies earlier in life. Who knows? What about other variables that may affect both mothers' ages at the births of their children and longer lifespans — variables such as financial status, marital stability, or other socio-economic factors? The women in this study were 33 years old during the Depression; this may have influenced both their life span and if or when they had children.

How do researchers handle confounding variables? The operative word is "control." They control for as many confounding variables that they can anticipate. In experiments involving human subjects, researchers have to battle many confounding variables. For example, in a study trying to determine the effect of different types and volumes of music on the amount of time grocery shoppers spend in the store (yes, they do think about that), researchers have to anticipate as many possible confounding variables ahead of time, and then control for them. What other factors besides volume and type of music could influence the amount of time you spend in a grocery store? I can think of several factors: gender, age, time of day, whether I have children with me, how much money I have, the day of the week, how clean and inviting the store is, how nice the employees are, and (most importantly) what my motive is — am I shopping for the whole week, or am I just running in to grab a candy bar?

How can researchers begin to control for so many possible confounding factors? Some of them can be controlled for in the design of the study, such as the time of the day, day of the week, and reason for shopping. But other factors (such as the perception of the store environment) depend totally on the individual in the study. The ultimate form of control for those person-specific confounding variables is to use pairs of people that are matched according to important variables, or to just use the same person twice: once with the treatment and once without. This type of experiment is called a *matched-pairs design.* (See Chapter 14 for more on this.)

Before believing any medical headlines (or any headlines for that matter), look to see how the study was conducted. Observational studies can't control for confounding variables, so their results are not as statistically meaningful (no matter what the statistics say) as the results of a well-designed experiment. In cases where an experiment can't be done (after all, no one can force you to have a baby after or before age 40), make sure the observational study is based on a large enough sample that represents a cross-section of the population.

Double-blinding the experiment

Well-designed experiments are done in a double-blind fashion. *Double-blind* means that neither the subjects nor the researchers know who got what treatment or who is in the control group. The research subjects need to be oblivious to which treatment they're getting so that the researchers can measure the placebo effect. But why shouldn't the researcher know who got what treatment? So that they don't treat subjects differently by either expecting (or not expecting) certain responses from certain groups. For example, if a researcher knows you're in the treatment group to study the side effects of a new drug, she may expect you to get sick and pay more attention to you than if she knew you were in the control group. This can result in biased data and misleading results.

If the researcher knows who got what treatment but the subjects don't know, the study is called a *blind* study (rather than a double-blind study). Blind studies are better than nothing, but double-blind studies are best. In case you're wondering: In a double-blind study, does *anyone* know which treatment was given to which subjects? Relax; typically a third party lab assistant does that part.

When analyzing an experiment, look to see whether it was a double-blind study. If not, the results can be biased.

Collecting good data

What constitutes "good" data? Statisticians use three criteria for evaluating data quality; each of the criteria really relates most strongly to the quality of the measurement instrument that's used in the process of collecting the data.

To decide whether or not you're looking at good data from an experiment, look for these characteristics:

✔ **Reliable (you can get repeatable results with subsequent measurements):** Many bathroom scales give unreliable data. You get on the scale, and it gives you one number. You don't believe the number, so you get off, get back on, and get a different number. (If the second number is lower, you'll most likely quit at this point; if not, you may continue getting on and off until you see a number you like.)

The point is, unreliable data come from unreliable measurement instruments. These instruments can go beyond scales to more intangible measurement instruments, like survey questions, which can give unreliable results if they're written in an ambiguous way (see Chapter 16 for more on this).

Find out how the data were collected when examining the results of an experiment. If the measurements are unreliable, the data could be inaccurate.

✔ **Unbiased (the data contain no systematic errors that either add to or subtract from the true values):** Biased data are data that systematically over measure or under measure the true result. Bias can occur almost anywhere during the design or implementation of a study. Bias can be caused by a bad measurement instrument (like a bathroom scale that's "always" five pounds over) by survey questions that lead participants in a certain way, or by researchers who know what treatment each subject received and have preconceived expectations.

Bias is probably the number-one problem in data. And worse yet, it can't really be measured. (For example, the margin of error doesn't measure bias. See Chapter 10 for details on the margin of error.) However, steps can be taken to minimize bias, as discussed in Chapter 16 and in the "Randomly assigning subjects to groups" section earlier in this chapter.

Be aware of the many ways that bias can come into play during the design or implementation of any study, and evaluate the study with an eye toward detecting bias. If a study contains a great deal of bias, you have to ignore the results.

✔ **Valid (the data measure what they're supposed to measure):** Checking the validity of data requires you to step back and look at the big picture. You have to ask the question: Do these data measure what they should be measuring? Or should the researchers have been collecting altogether different data? The appropriateness of the measurement instrument used is important. For example, asking students to report their high school math grades may not be a valid measure of actual grades. A more valid measure would be to look at each student's transcript. Measuring the prevalence of crime by the number of crimes is not valid; the *crime rate* (number of crimes per capita) should be used.

Before accepting the results of an experiment, find out what data were measured and how they were measured. Be sure the researchers are collecting data that are appropriate for the goals of the study.

Analyzing the data properly

After the data have been collected, they're put into that mysterious box called the *statistical analysis*. The choice of analysis is just as important (in terms of the quality of the results) as any other aspect of a study. A proper analysis should be planned in advance, during the design phase of the experiment. That way, after the data are collected, you won't run into any major problems during the analysis.

Here's the bottom line when selecting the proper analysis. Ask yourself the question, "After the data are analyzed, will I be able to answer the question that I set out to answer?" If the answer is "no," that analysis isn't appropriate.

The basic types of statistical analyses include confidence intervals (used when you're trying to estimate a population value, or the difference between two population values); hypothesis tests (used when you want to test a claim about one or two populations, such as the claim that one drug is more effective than another); and correlation and regression analyses (used when you want to show if and/or how one variable can predict or cause changes in another variable). See Chapters 13, 15, and 18, respectively, for more on each of these types of analyses.

When choosing how you're going to analyze your data, you have to make sure that the data and your analysis will be compatible. For example, if you want to compare a treatment group to a control group in terms of the amount of weight lost on a new (versus an existing) diet program, you need to collect data on how much weight each person lost (not just each person's weight at the end of the study).

Drawing appropriate conclusions

In my opinion, the biggest mistakes researchers make when drawing conclusions about their studies are:

- ✔ Overstating their results
- ✔ Making connections or giving explanations that aren't backed up by the statistics
- ✔ Going beyond the scope of the study in terms of who the results apply to

Each of these problems is discussed in the following sections.

Overstating the results

Many times, the headlines in the media will overstate actual research results. When you read a headline or otherwise hear about a study, be sure to look further to find out the details of how the study was done and exactly what the conclusions were.

Press releases often overstate results, too. For example, in a recent press release by the National Institute for Drug Abuse, the researchers claimed that Ecstasy use was down in 2002 from 2001. However, when you look at the actual statistical results in the report, you find that the percentage of teens in the sample who said they'd used Ecstasy was lower in 2002 than it was in 2001, but this difference was not found to be statistically significant (even though many other results were found to be statistically significant). This means that although fewer teens in the sample used Ecstasy in 2002, the difference wasn't enough to account for more than chance variability from sample to sample. (See Chapter 14 for more about statistical significance.)

Headlines and leading paragraphs in press releases and newspaper articles often overstate the actual results of a study. Big results, spectacular findings, and major breakthroughs are what make the news these days, and reporters and others in the media are constantly pushing the envelope in terms of what is and is not newsworthy. How can you sort out the truth from exaggeration? The best thing to do is to read the fine print.

Taking the results one step beyond the actual data

A study that links having children later in life to longer lifespans illustrates another point about research results. Do the results of this observational study mean that having a baby later in life can make you live longer? "No," said the researchers. Their explanation of the results was that having a baby later in life may be due to women having a "slower" biological clock, which presumably would then result in the aging process being slowed down.

My question to these researchers is, "Then why didn't you study *that,* instead of just looking at their ages?" I don't see any data in this study that would lead me to conclude that women who had children after age 40 aged at a slower rate than other women, so in my view, the researchers shouldn't make that conclusion yet. Or the researchers should state clearly that this view is only a theory and requires further study. Based on the data in this study, the researchers' theory seems like a leap of faith (although as a 41-year-old new mom, I'll hope for the best!).

Frequently, in a press release or news article, the researcher will give an explanation about *why* he thinks the results of the study turned out the way they did and what implications these results have for society as a whole. (These explanations may have been in response to a reporter's questions about the research, questions that were later edited out of the story, leaving only the juicy quotes from the researcher.) Many of these after-the-fact explanations are no more than theories that have yet to be tested. In such cases, you should be wary of conclusions, explanations, or links drawn by researchers that aren't backed up by their studies.

Generalizing results to people beyond the scope of the study

You can make conclusions only about the population that's represented by your sample. If you sample men only, you can't make conclusions about women. If you sample healthy young people, you can't make your conclusions about everyone. But many researchers try to do just that. This is a common practice that can give misleading results. Watch out for this one!

Here's how you can determine whether a researcher's conclusions measure up:

- ✔ Find out what the target population is (that is, the group that the researcher wants to make conclusions about).

- ✔ Find out how the sample was selected and see whether the sample is representative of that target population (and not some more narrowly defined population).

- ✔ Check the conclusions made by the researchers; make sure they're not trying to apply their results to a broader population than they actually studied.

Making Informed Decisions about Experiments

Just because someone says they conducted a "scientific study" or a "scientific experiment," doesn't mean it was done right or that the results are credible. Unfortunately, I've come across a lot of bad experiments in my days as a statistical consultant. The worst part is that if an experiment was done poorly, you can't do anything about it after the fact except ignore the results — and that's exactly what you need to do.

Here are some tips that help you make an informed decision about whether to believe the results of an experiment, especially one whose results are very important to you.

- ✔ When you first hear or see the result, grab a pencil and write down as much as you can about what you heard or read, where you heard or read it, who did the research, and what the main results were. (I keep pencil and paper in my TV room and in my purse just for this purpose.)

- ✔ Follow up on your sources until you find the person who did the original research, and then ask them for a copy of the report or paper.

- ✔ Go through the report and evaluate the experiment according to the eight steps for a good experiment described in the "Designing a Good Experiment" section of this chapter. (You really don't have to understand everything written in a report in order to do that.)

- ✔ Carefully scrutinize the conclusions that the researcher makes regarding his or her findings. Many researchers tend to overstate results, make conclusions beyond the statistical evidence, or try to apply their results to a broader population than the one they studied.

- ✔ Never be afraid to ask questions of the media, the researchers, and even your own experts. For example, if you have a question about a medical study, ask your doctor. He or she will be glad that you're an empowered and well-informed patient!

Chapter 18

Looking for Links: Correlations and Associations

In This Chapter

▶ Understanding the statistics of relationships

▶ Distinguishing between association, correlation, and causation

▶ Making predictions based on known relationships

▶ Detecting misleading results

Everyone seems to want to tell you about the latest relationships, correlations, associations, or links they've found. Many of these links come from medical research, as you may expect. The job of medical researchers is to tell you what you should and shouldn't be doing in order to live a longer and healthier life.

Here are some recent news releases provided by the National Institutes of Health (NIH):

✔ Sedentary activities (like TV watching) are associated with an increase in obesity and an increase in the risk of diabetes in women.

✔ Anger expression may be *inversely related* to the risk of heart attack and stroke. (Those who express anger have a decreased risk.)

✔ Light-to-moderate drinking reduces the risk of heart disease in men.

✔ Immediate treatment helps delay the progression of glaucoma.

Reporters love to tell people about the latest links, because these stories can make big news. Some of the recommendations seem to change from time to time, though; for example, one minute, zinc is recommended to prevent colds, and the next minute it's "out in the cold." Many of the relationships you see in the media are touted as cause-and-effect relationships, but can you believe these reports? (For example, does having her first baby later in life cause a woman to live longer?) Are you so skeptical that you just don't believe much of anything anymore?

If you're a confused information consumer when you hear about links and correlations, take heart; this chapter can help. You discover what it truly means for two factors to be correlated, associated, or have a cause-and-effect relationship, and when and how to make predictions based on those relationships. You also gain the skills to dissect and evaluate research claims and to make your own decisions about those headlines and sound bites alerting you to the latest correlation.

Picturing the Relationship: Plots and Charts

An article in *Garden Gate* magazine caught my eye. "Count Cricket Chirps to Gauge Temperature," the title read. According to the article, all you have to do is find a cricket, count the number of times it chirps in 15 seconds, add 40, and voila! You've just predicted the temperature in degrees Fahrenheit.

The National Weather Service Forecast Office even puts out its own Cricket Chirp Converter. You enter the number of times the cricket chirps in 15 seconds, and the converter gives you the temperature estimated in four different units, including degrees Fahrenheit and Celsius.

A fair amount of research does support the claim that the frequency of cricket chirps is related to temperature. For illustration, I've taken a subset of some of the data and presented it in Table 18-1. Notice that each observation is composed of two variables that are tied together, in this case number of times the cricket chirped in 15 seconds, and the temperature at that time (in degrees Fahrenheit). Statisticians call this type of two-dimensional data *bivariate* data. Each observation contains one pair of data collected simultaneously.

Table 18-1	Cricket Chirps and Temperature Data (Excerpt)
Number of Chirps in 15 Seconds	**Temperature (in Degrees Fahrenheit)**
18	57
20	60
21	64
23	65
27	68
30	71

Number of Chirps in 15 Seconds	Temperature (in Degrees Fahrenheit)
34	74
39	77

A recent press release put out by the Ohio State University Medical Center also caught my attention. The headline says that aspirin can prevent polyps in colon cancer patients. Having had a close relative who succumbed to this disease, I was heartened at the prospect that researchers are making progress in this area and decided to look into it. The data from the aspirin versus colon polyps study are summarized in Table 18-2.

Table 18-2   Summary of Aspirin versus Polyps Study Results

Group	% Developing Polyps*
Aspirin	17
Non-aspirin (placebo)	27

*total sample size = 635 (approximately half were randomly assigned to each group)

The raw data for this study contain 635 lines. Each line represents a person in the study and includes the person's identification number, the group to which he or she was assigned (aspirin or non-aspirin), and whether or not the subject developed polyps during the study period (yes or no). For example, line 1 of this data set may look like this:

    ID#22292   GROUP = ASPIRIN   DEVELOPED POLYPS = NO

If you look at the raw bivariate data for this large data set, you'd probably have a hard time deducing any relationship between the variables: 635 lines of data would be hard for anyone (except a computer) to make any sense out of. For the cricket chirps versus temperature data, even if you can see a general pattern in the raw data (for example, noticing that as the number of chirps increases the temperature seems to also increase), the exact relationship is hard to pinpoint.

To make sense out of any data, you should first organize them using a table, chart, or graph (see Chapter 4). When the data are bivariate and you're looking for links between the two variables, the charts and graphs need to have two dimensions to them as well, just like the data do. That's the only way you can explore possible connections between the variables.

Displaying bivariate numerical data

In the case where both variables are quantitative (that is, numerical, such as measures of height and weight), the bivariate data are typically organized in a graph that statisticians call a *scatterplot*. A scatterplot has two dimensions, a horizontal dimension (called the *x*-axis) and a vertical dimension (called the *y*-axis). Both axes are numerical; each one contains a number line.

Making a scatterplot

Placing observations (or points) on a scatterplot is similar to finding a city on a map that uses letters and numbers to mark off sections of the map. Each observation has two coordinates; the first corresponds to the first piece of data in the pair (that's the *x*-coordinate, the amount that you go left or right). The second coordinate corresponds to the second piece of data in the pair (that's the *y*-coordinate, the amount that you go up or down). Intersect the two coordinates, and that's where you place the point representing that observation. Figure 18-1 shows a scatterplot for the cricket chirps versus temperature data listed in Table 18-1. Because I put the data in order according to their *x*-values (number of chirps) when I generated Table 18-1, the points on the scatterplot (going from left to right) correspond in order to the observations given in Table 18-1.

Figure 18-1:
Scatterplot
of cricket
chirps
versus
outdoor
temperature.

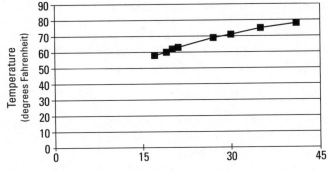

Number of Cricket Chirps per 15 Seconds

Interpreting a scatterplot

You interpret a scatterplot by looking for trends in the data as you go from left to right:

✔ If the data resemble an uphill line as you move from left to right, this indicates a *positive linear* (or *proportional*) *relationship*. As *x* increases (moves right one unit), *y* increases (moves up) a certain amount.

✔ If the data resemble a downhill line as you move from left to right, this indicates a *negative linear* (or *inverse*) *relationship*. That means that as *x* increases (moves right one unit), *y* decreases (moves down) a certain amount.

✔ If the data don't seem to resemble any kind of line (even a vague one), this means that no linear relationship exists.

Looking at Figure 18-1, there does appear to be a positive linear relationship between number of cricket chirps and the temperature. That is, as the cricket chirps increase, the temperature increases, as well. Is this a cause-and-effect relationship (in other words, do temperature increases *cause* crickets to chirp faster)? That remains to be seen, because these data come only from an observational study, not an experiment (see Chapter 17).

Visual displays of bivariate data show possible associations or relationships between two variables. However, just because your graph or chart shows that something is going on, it doesn't mean that a cause-and-effect relationship exists. For example, if you look at a scatterplot of ice cream consumption and murder rates, those two variables have a positive linear relationship, too. Yet, no one would claim that ice cream consumption causes murders, or that murder rates affect ice cream consumption. If someone is trying to establish a cause-and-effect relationship by showing a chart or graph, dig deeper to find out how the study was designed and how the data were collected, and then evaluate the study appropriately using the criteria outlined in Chapter 17.

Other types of trends may exist besides the uphill or downhill linear trends. Variables can be related to each other through a curved relationship or through various exponential relationships; these are beyond the scope of this book. The good news, however, is that many relationships can be characterized as uphill or downhill linear relationships.

Displaying bivariate categorical data

In the case where both variables are categorical (such as the gender of the respondent and whether the respondent does or does not support the president), the data are typically summarized in what statisticians call a *two-way table* (meaning a table that has rows representing the categories of the first variable and columns representing the categories of the second variable).

For example, in the aspirin versus colon polyps study, the two variables were categorical: whether or not the colon-cancer patient took aspirin (yes or no), and whether or not that person developed any more polyps (yes or no). Note that Table 18-2 in the "Picturing the Relationship: Plots and Charts" section is a two-way table.

A more visually pleasing way to organize two-dimensional data is to use a bar graph or a series of pie charts. Figure 18-2 shows a bar graph that indicates the percentage of patients who developed polyps for the aspirin group compared to the non-aspirin (placebo) group. Figure 18-3 shows two pie charts, one for the aspirin group and one for the non-aspirin group. Each pie chart shows the percentage in that group that developed polyps.

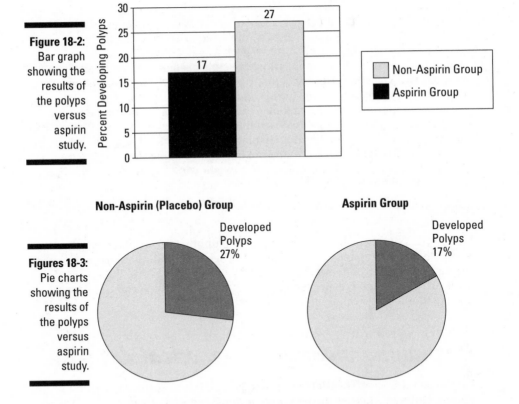

Figure 18-2: Bar graph showing the results of the polyps versus aspirin study.

Figures 18-3: Pie charts showing the results of the polyps versus aspirin study.

Because the bars in the bar graph have very different sizes and because the two pie charts look quite different, a relationship does appear to exist between aspirin-taking and the development of a polyp among the subjects of this study. (The operative word here is "appears." Hypothesis tests for two proportions need to be done in order to be sure that these differences in the samples can safely apply to their respective populations. See Chapter 15 for more on this.)

Be skeptical of anyone who draws conclusions about a relationship between two variables by only showing a chart or graph. Looks can be deceiving (see Chapter 4). Statistical measurements and tests should be done to show that such relationships are statistically meaningful (see Chapter 14).

In the case of the aspirin versus polyp data, you may think that the second variable (polyps) is numerical because the value is represented by a percentage. But, in fact, that's not the case. Percentages are just a handy way to summarize the data from a categorical variable. In this case, the second variable is whether or not polyps developed (yes/no) for each patient. (This is a categorical variable.) The percentages simply summarize all of the patients in each yes and no category.

Quantifying the Relationship: Correlations and Other Measures

After the bivariate data have been organized, the next step is to do some statistics that can quantify or measure the extent and nature of the relationship.

Quantifying a relationship between two numerical variables

If both variables are numerical or quantitative, statisticians can measure the direction and the strength of the linear relationship between the two variables x and y. Data that "resemble an uphill line" have a positive linear relationship, but it may not necessarily be a strong relationship. The strength of the relationship depends on how closely the data resemble a straight line. Of course, varying levels of "closeness to a line" exist. Plus you have to distinguish between the positive and the negative relationships. Can one statistic measure all of that? Sure!

Statisticians use what they call the *correlation coefficient* to measure the strength and direction of the linear relationship between x and y.

Calculating the correlation coefficient (r)

The formula for the correlation coefficient (denoted r) is

$$r = \frac{1}{n-1} \sum \frac{(x - \overline{x})(y - \overline{y})}{s_x s_y}.$$

To calculate the correlation coefficient:

1. **Find the mean of all the x values (call it $\bar{x}$) and the mean of all the y values (call it $\bar{y}$).**

   See Chapter 5 for calculations.

2. **Find the standard deviation of all the x values (call it s_x) and the standard deviation of all the y values (call it s_y).**

   See Chapter 5.

3. **For each (x, y) pair in the data set, take x minus $\bar{x}$ and y minus $\bar{y}$, and then multiply these differences together.**

4. **Add all of these products together to get a sum.**

5. **Divide the sum by $s_x \times s_y$.**

6. **Divide that result by $n - 1$, where n is the number of (x, y) pairs.**

For example, suppose you have the data set (3, 2), (3, 3), and (6, 4). Following the preceding steps, you can calculate the correlation coefficient. Note that the x values are 3, 3, and 6, and the y values are 2, 3, and 4.

1. **$\bar{x}$ is $12 \div 3 = 4$, and $\bar{y}$ is $9 \div 3 = 3$.**

2. **The standard deviations are $s_x = 1.73$ and $s_y = 1.00$.**

3. **The differences found in Step 3 multiplied together are: $(3 - 4)(2 - 3) = (-1)(-1) = 1$; $(3 - 4)(3 - 3) = (-1)(0) = 0$; $(6 - 4)(4 - 3) = (+2)(+1) = +2$.**

4. **The results from Step 3, all added up, are $1 + 0 + 2 = 3$.**

5. **Dividing the Step 4 result by $s_x \times s_y$ gives you $3 \div (1.73 \times 1.00) = 3 \div 1.73 = 1.73$.**

6. **Dividing the Step 5 result by $3 - 1$ (which is 2), you get 0.87.**

   This is the correlation.

Interpreting the correlation

The correlation r is always between -1 and $+1$.

- ✔ A correlation of exactly -1 indicates a perfect downhill linear relationship.

- ✔ A correlation close to -1 indicates a strong downhill linear relationship.

- ✔ A correlation close to 0 means that no linear relationship exists.

- ✔ A correlation close to $+1$ indicates a strong uphill linear relationship.

- ✔ A correlation of exactly $+1$ indicates a perfect uphill linear relationship.

Many folks make the mistake of thinking that a correlation of −1 is a bad thing, indicating no relationship. In fact, the opposite is true! A correlation of −1 means that the data are lined up in a perfect straight line, the strongest linear relationship you can get. That line just happens to be going downhill — that's what the minus sign is for!

How "close" do you have to get to −1 or +1 to indicate a strong linear relationship? Most statisticians like to see correlations above +0.6 (or below −0.6) before getting too excited about them. However, don't expect a correlation to always be +0.99 or −0.99; these are real data, and real data aren't perfect.

Figure 18-4 shows examples of what various correlations look like, in terms of the strength and direction of the relationship.

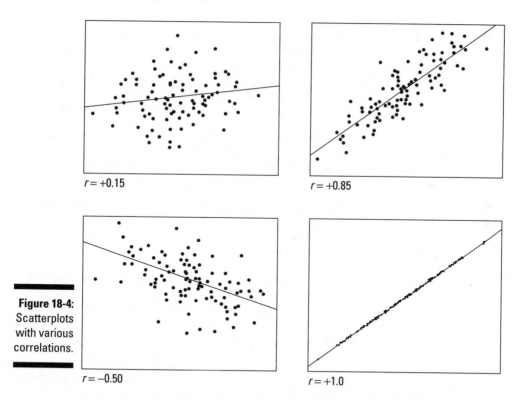

$r = +0.15$

$r = +0.85$

$r = -0.50$

$r = +1.0$

Figure 18-4:
Scatterplots
with various
correlations.

For my subset of the cricket chirps versus temperature data, I calculated a correlation of 0.98, which is almost unheard of in the real world (these crickets are *good!*).

Understanding the properties of the correlation coefficient

Here are some useful properties of correlations:

- ✔ The correlation is a unitless measure. This means that if you change the units of x or y, the correlation won't change. (For example, changing from Fahrenheit to Celsius won't affect the correlation between the frequency of chirps and the outside temperature.)

- ✔ The values of x and y can be switched in the data set, and the correlation won't change.

Quantifying a relationship between two categorical variables

If both variables are categorical (such as whether or not the patient took aspirin and whether or not the patient developed polyps), you really can't use the word "correlation" to describe the relationship, because correlation measures the strength of the linear relationship between numerical variables. (This mistake occurs in the media all the time, and it drives statisticians crazy!)

The word that is used to describe a relationship between two categorical variables is *association*. Two categorical variables (such as treatment group and outcome) are associated if the percentage of subjects who had a certain outcome in one group is significantly different than the percentage who had that same outcome in the other group. In the aspirin versus polyp example discussed in the first section of this chapter, researchers found that in the aspirin group, 17% of the colon cancer patients developed polyps, whereas in the placebo group, 27% developed polyps. Because these percentages are quite different, the two variables are associated.

How different do the percentages have to be in order to signal a meaningful association between the two variables? The difference found by the sample has to be *statistically significant*. That way, the same conclusion about a relationship can be made about the whole population, not just for a particular data set. A hypothesis test of two proportions will work for this purpose (see Chapter 15 for details on this test). I analyzed the data from the aspirin versus polyps study using that test and got a p-value of less than 0.0001. That means these results are highly significant. (See Chapter 14 for more on p-values.) You can see why the researchers stopped this study midstream and decided to give everyone the aspirin treatment!

Explaining the Relationship: Association and Correlation versus Causation

If two variables are found to be either associated or correlated, that doesn't necessarily mean that a cause-and-effect relationship exists between the two variables. Whether two variables are found to be causally associated depends on how the study was conducted. Only a well-designed experiment (see Chapter 17) or several different observational studies can show that an association or a correlation crosses over into a cause-and-effect relationship.

Taking aspirin does seem to help

I feel confident about the conclusions drawn by the researchers in the aspirin versus polyps study discussed in the first section of this chapter; this study was a well-designed experiment, according to the criteria established in Chapter 17. It included random assignment of patients to treatments, it had large enough sample sizes to obtain accurate information, and it controlled for confounding variables. This means that the researchers truly are entitled to the headline of the press release, "Aspirin Prevents Polyps in Colon Cancer Patients." Because of the design of this study, you can say that a cause-and-effect relationship (association) exists between whether the colon cancer patients took aspirin on a daily basis and whether polyps developed.

Turning up the heat on cricket chirps

Does the outside temperature cause crickets to chirp faster or slower? (Obviously the reverse isn't true, but what about the possible causation in this direction?) Some people speculate that changes in the outside temperature cause crickets to chirp at different frequencies. However, I'm not aware of any data based on experiments (as opposed to observational studies) that would confirm or deny a cause-and-effect relationship here. Perhaps you can do an experiment of your own and turn up the heat on some crickets and see what happens! (Before leaping into this — yes, the pun was intended — be sure to design a good experiment following the criteria in Chapter 17.)

Making Predictions: Regression and Other Methods

After you've found a relationship between two variables and you have some way of quantifying this relationship, you can create a model that allows you to use one variable to predict another.

Making predictions with correlated data

In the case of two numerical variables, if a strong correlation has been established, researchers often use the relationship between x and y to make predictions. Because x is correlated with y, a *linear relationship* exists between them. This means that you can describe the relationship with a straight line. If you know the slope and the y-intercept of that line, then you can plug in a value for x and predict the average value for y. In other words, you can predict y from x.

Because the correlation between cricket chirps and temperature is so high ($r = 0.98$), you can find a line that fits the data. This means that you want to find the one line that best fits the data (in terms of the average distance from all of the points in the data set to the line you generate). Statisticians call this search for the best-fitting line performing a *regression analysis*.

Never do a regression analysis unless you've already found a strong correlation (either positive or negative) between the two variables. I've seen cases where researchers go ahead and make predictions when a correlation was as low as 0.2! That doesn't make any sense. If the scatterplot of the data doesn't resemble a line to begin with, you shouldn't try to use a line to fit the data and to make predictions about the population.

Before examining any model that predicts one variable from another, find the correlation first; if the correlation is too weak, stop right there.

You may be thinking that you have to try lots and lots of different lines to see which one fits best. Fortunately, this is not the case (although eyeballing a line on the scatterplot does help you think about what you'd expect the answer to be). The best-fitting line has a distinct slope and y-intercept that can be calculated using formulas (and, I may add, these formulas aren't too hard to calculate).

Getting a formula for best-fitting line

The formula for the *best-fitting line* (or *regression line*) is $y = mx + b$, where m is the slope of the line and b is the y-intercept. The *slope* of a line is the change in y over the change in x. For example, a slope of $10/3$ means that as x moves to the right 3 units, the y-value moves up 10 units, as you move from one point on the line to the next. The y-intercept is that place on the y-axis where the line crosses. For example, in the equation $y = \frac{10}{3}x - 6$, the line crosses the y-axis at the point -6. The coordinates of this point are $(0,-6)$ — because you are crossing the y-axis, the x value of the y-intercept is always 0. To come up with the best-fitting line, you need to find values for m and b so that you have a real equation of a line (for example, $y = 2x + 3$; or $y = -10x - 45$).

To save a great deal of time calculating the best-fitting line, keep in mind that five well-known summary statistics are all you need to do all the necessary calculations. Statisticians call them the *big-five summary statistics:*

✔ The mean of the x values (denoted $\bar{x}$)

✔ The mean of the y values (denoted $\bar{y}$)

✔ The standard deviation of the x values (denoted s_x)

✔ The standard deviation of the y values (denoted s_y)

✔ The correlation between x and y (denoted r)

(This chapter and Chapter 5 contain formulas and step-by-step instructions for these statistics.)

Finding the slope of the best-fitting line

The formula for the slope, m, of the best-fitting line is $m = r\left(\frac{s_y}{s_x}\right)$, where r is the correlation between x and y, and s_y and s_x are the standard deviations of the y-values and the x-values, respectively (see Chapter 5 for more on standard deviation).

To calculate the slope, m, of the best-fitting line:

1. **Divide s_y by s_x.**

2. **Multiply the result in Step 1 by r.**

The slope of the best-fitting line can be a negative number because the correlation can be a negative number. A negative slope indicates that the line is going downhill.

The formula for slope simply takes the correlation (a unitless measurement) and attaches units to it. Think of $s_y \div s_x$ as the change in y over the change in x. And the standard deviations are each in terms of their original units (for example, temperature in Fahrenheit and number of cricket chirps in 15 seconds).

Finding the y-intercept of the best-fitting line

The formula for the y-intercept, b, of the best-fitting line is $b = \bar{y} - m\bar{x}$, where $\bar{y}$ and $\bar{x}$ are the means of the y-values and the x-values, respectively, and m is the slope (the formula for which is given in the preceding section).

To calculate the y-intercept, b, of the best-fitting line:

1. **Find the slope, m, of the best-fitting line using the steps listed in the preceding section.**

2. **Multiply $m \times \bar{x}$.**

3. **Take $\bar{y}$ and subtract your result from Step 2.**

Always calculate the slope before calculating the *y*-intercept. The formula for the *y*-intercept contains the slope in it, so you need *m* to calculate *b*.

Finding the best-fitting line for cricket chirps and temperature

Although the formula for the line that best fits the relationship between cricket chirps and temperature is subject to a bit of discussion (see Appendix), the consensus seems to be that a good working model for this relationship is $y = x + 40$, or temperature = $1 \times$ (number of chirps in 15 seconds) + 40, where the temperature is in degrees Fahrenheit. Note that the slope of this line is 1, x = number of chirps in 15 seconds, and y = temperature in degrees Fahrenheit.

Notice that the formulas for the slope and *y*-intercept are in the form of *x* and *y*, so you have to decide which of your two variables you'll call *x* and which you'll call *y*. When doing correlations, the choice of which variable is *x* and which is *y* doesn't matter, as long as you're consistent for all the data; but when fitting lines and making predictions, the choice of *x* and *y* does make a difference. Take a look at the preceding formulas — switching the roles of *x* and *y* makes all of the formulas change.

So how do you determine which variable is which? In general, *x* is the variable that is the predictor. Statisticians call *x* the *explanatory variable,* because if you change *x*, that explains why and how *y* is going to change. In this case, *x* is the number of cricket chirps in 15 seconds. The *y* variable is called the *response variable;* it responds to changes in *x*. In other words, *y* is being predicted by *x*. Here, *y* is the temperature.

Comparing the working model to the data subset

The big-five summary statistics from the subset of cricket data are shown in Table 18-3.

Table 18-3	Cricket Data Big-Five Summary Statistics		
Variable	*Mean*	*Standard Deviation*	*Correlation*
# Chirps *(x)*	$\bar{x} = 26.5$	$s_x = 7.4$	$r = +0.98$
Temp *(y)*	$\bar{y} = 67$	$s_y = 6.8$	

The slope, *m,* for the best-fitting line for the subset of cricket chirp versus temperature data is $r \times (s_y \div s_x) = 0.98 \times (6.8 \div 7.4) = 0.98 \times 0.919 = 0.90$. Now, to find the *y*-intercept, *b,* you take $\bar{y} - m \times \bar{x}$, or $67 - (0.90)(26.5) = 67 - 23.85 = 43.15$. So the best-fitting line for predicting temperature from cricket chirps based on the data is: $y = 0.9x + 43.2$, or

temperature (in degrees Fahrenheit) = $0.9 \times$ (number of chirps in 15 seconds) + 43.2.

Note that the preceding equation is close to, but not quite the same as, the working model: $y = x + 40$. Why isn't the preceding equation exactly the same as the working model? A couple of reasons come to mind. First, "working model" is fancy language for "not necessarily precise, but very practical." I'm guessing that over the years, the slope has been rounded to the nearest whole number (1) and the y-intercept has been rounded to the nearest ten (which makes it 40) just to make it easier for people to remember and more fun to write about. (This isn't good statistical practice and is an example of how statistics can drift over a period of years.) Second, the data I used are just a random subset of the original data set (for purposes of illustration) and will be a bit off, just by chance (see Chapter 9 for more on variation between samples). However, because the data are so highly correlated, the difference between one sample of data and another shouldn't be much.

Predicting temperature with cricket chirps

The best-fitting line for predicting temperature with cricket chirps with my subset of data was found to be $y = 0.9x + 43.2$. Any equation or function that is used to estimate or predict a relationship between two variables is called a *statistical model*. Using this model, you can predict temperature using cricket chirps. How do you do it? Choose a relevant value for x, plug it into the model, and find the expected value for y.

For example, if you want to predict the temperature and you know that the cricket in your backyard chirped 35 times in 15 seconds, you plug in 35 for x and find out what y is. Here goes: $y = 0.9(35) + 43.2 = 31.5 + 43.2 = 74.7$. So, knowing that the cricket chirped 35 times in 15 seconds, you can predict that the expected temperature is about 75 degrees Fahrenheit.

Just because you have a model doesn't mean you can plug in *any* value for x and do a good job of predicting y. For example, you can't plug in a number higher than 39 or lower than 18. Why not? Because, for this example, you don't have any data for x in those ranges (see Table 18-1). Who's to say the line still works outside of the area where data were collected? Do you really think that as the temperature rises higher and higher into infinity the crickets will chirp faster and faster, without limit? At some point the poor crickets would die of heat exposure. Similarly, crickets won't survive in extremely cold weather, so you can't plug in extremely low values of x and expect the model to work.

Making predictions using x values that fall outside the range of your data is a no-no. Statisticians call this *extrapolation;* watch for researchers who try to make claims beyond the range of their results.

Because the best-fitting line is a model describing the overall relationship between x and y, you're not really predicting y, you're predicting the *expected* (or average) value of y for any given x.

Making predictions with two associated categorical variables

After two categorical variables have been found to be associated, you can make predictions (estimates) about the percentage in each group regarding one of the variables, or you can estimate the difference between the percentages in the two groups. In both cases, you use confidence intervals to make those estimates (see Chapters 12 and 13).

With the aspirin versus colon polyps example given in the first section of this chapter, among the group taking aspirin, the percentage of colon cancer patients who developed polyps was 17%, compared to 27% in the non-aspirin group (refer to Table 18-2). The prediction you can make here is the following: If you're a colon cancer patient, you're less likely to develop a subsequent colon polyp if you take 325 mg of aspirin every day.

You can get more specific about this prediction. You can estimate the chance of a colon cancer patient developing a polyp if he or she takes aspirin every day by finding a confidence interval. Because 17% of the sample of aspirin takers developed polyps, this means that the chance that any individual in that population (that is, colon cancer patients) will develop subsequent polyps if they take aspirin every day is 0.17, plus or minus the margin of error, or in this case, plus or minus 0.04. In other words, for a colon cancer patient taking 325 mg of aspirin daily, the chance of developing subsequent polyps is anywhere from 17% − 4% to 17% + 4%, or from 13% to 21%. (See Chapter 13 for the formula for a confidence interval for a single population proportion, p, and Chapter 10 for finding the margin of error.)

Another way to make predictions in this situation is to estimate the decreased risk of developing polyps, should the patients take an aspirin every day. This can be accomplished by calculating a confidence interval for the difference between the two proportions, where p_1 is the proportion of patients in the control group who developed polyps, and p_2 is the proportion of patients in the aspirin group who developed polyps. The difference you're interested in here is $p_1 - p_2$. This confidence interval is 0.27 − 0.17 = 0.10, plus or minus the margin of error, or, in this case, plus or minus 0.03. So, if you're a colon cancer patient taking aspirin each day, your risk of developing subsequent polyps decreases anywhere from 10% − 3% to 10% + 3%, or from 7% to 13%. (See Chapter 13 for the formula for a confidence interval for the difference between two population proportions.)

Chapter 19

Statistics and Toothpaste: Quality Control

. . . .

In This Chapter

▶ Seeing how statistics helps improve products

▶ Keeping within the specs: quality control basics

▶ Monitoring the process

. . . .

The most successful manufacturing companies are all about quality control. They want you, as a customer, to be satisfied with their products and buy them again and again. They want you to be so pleased that you'll tell your friends, neighbors, coworkers, and even people on the street how wonderful their products are. How do companies ensure that you're going to be satisfied with their products? One criterion for customer satisfaction is product quality, and believe it or not, the field of statistics plays a vital role in the assessment of product quality and in quality improvement. This chapter shows you how.

Full-Filling Expectations

Customers expect products to fulfill their expectations, and one expectation is that a package contains the amount of product that was promised. Another expectation is a certain level of consistency each time the product is purchased. How full do you expect a bag of potato chips to be? Doesn't it seem strange that an 8-ounce bag of potato chips can look so large yet actually contain so few potato chips? (Manufacturers say they insert air into the package before sealing it in order to protect the product from damage.) If the package labeling indicates that the weight of the package is 8 ounces, and the package contains 8 ounces, you really can't complain. But will you feel slighted if the bag turns out to be underfilled?

Suppose the package says it contains 8 ounces but it actually contains only 7.8 ounces; will you get upset? You probably wouldn't even notice that amount of difference. But what if the bag contains only 6 ounces of chips? How about 4 ounces? At some point, you're going to notice. And what will your reaction be? You may:

✔ Just let it slide (unless the problem happens again and again).

✔ Return the product to the store and demand a refund.

✔ Write a letter to the company and complain.

✔ Decide not to buy the product again.

✔ Complain to the Better Business Bureau (BBB) or a government agency.

✔ Organize a boycott of the product.

✔ Try to get a job with the company to be "part of the solution instead of part of the problem."

Now, some of these options may seem a bit over the top, especially if you got cheated out of only a couple of handfuls of potato chips. But suppose you bought a new car that turned out to be a lemon, your child almost choked on a part that came loose in his crib, or you got sick from some hamburger that you bought at the store yesterday. Quality can be a critical and serious issue. Although standards for many consumer products are developed and enforced by the U.S. government (for example, the Food and Drug Administration), problems do arise from time to time in the manufacturing process. Here are just a few of the factors that can affect the quality of a product during production:

✔ The employees perform inconsistently (due to differences in skill levels, training, and working conditions, or due to the effect of shift changes, poor morale, human error, and so on).

✔ The managers and/or supervisors are inconsistent or unclear about expectations and/or their responses to problems that arise.

✔ The production equipment performs inconsistently (due to the equipment not being sufficiently maintained, parts wearing out, machines breaking down or malfunctioning, or simply because of differences between individual machines or manufacturing lines).

✔ The machines and equipment aren't designed to be precise or sensitive enough.

✔ The raw materials used in the manufacturing process are inconsistent.

✔ The environment (temperature, humidity, air purity, and so on) isn't consistently controlled.

✔ The monitoring process is insufficient or inefficient.

Spurred on by the need to please customers and follow government regulations, successful manufacturers are always looking for ways to improve the quality of their products. One popular phrase used by the manufacturing industry is total quality management, or TQM. *Total quality management* focuses on developing ways to continually monitor, assess, and improve the manufacturing process from the beginning of the process to the end. TQM was popularized in the United States by a famous and beloved statistician, Dr. W. Edwards Deming, who developed a nationally known and often-used list called "14 Points for Management." Deming's philosophy was that if you build quality into the product in the first place, you'll lower costs, increase productivity, and be more competitive.

How do statistics factor into product quality? Statistics are used to determine and set specifications and to monitor all aspects of the manufacturing process to ensure that those specifications are met. Statistics are used to help decide when a process needs to be stopped, as well as to identify problems before they occur. In an overall sense, statistical data provide feedback (often continuously) to the manufacturer about product quality as part of the total quality management philosophy. The role of statistics in monitoring and improving product quality throughout the manufacturing process is called *statistical process control* (SPC). The subject of statistical process control can fill up an entire book by itself; however, you can get a very good sense of how statistics factor into quality control by understanding and applying some of the basic ideas discussed in the following sections.

Squeezing Quality out of a Toothpaste Tube

Although consumers appear to have reached a level of acceptance that potato chip bags will always be (or at least seem to be) woefully underfilled, they still hold toothpaste tubes to a higher standard, expecting them to be consistently filled to the top. (Toothpaste manufacturers must know how hard you struggle to squeeze that last bit of toothpaste out of the tube.)

Thankfully, the tube-filling industry (which includes its own Tube Council) rises to the occasion and takes the whole tube-filling concept very seriously. You can even find "Tube-Filling Frequently Asked Questions" on a Web site set up by one of the companies specializing in this area. One of these frequently asked questions addresses the issue of how quality is ensured in tube filling.

The goals of the tube-filling equipment, according to the industry, are accuracy and consistency. The dosing mechanism is the key to achieving these goals. (*Dosing mechanism* is the industry's lingo for the machine that actually fills the tubes.) The following are some important features of toothpaste-tube-dosing equipment:

✔ A mechanism for properly cutting off the flow, to eliminate drip or stringing

✔ A system to eliminate air in the filling process

✔ A mechanism that stops the machine from trying to fill tubes that, for some reason, are missing

✔ A system that's designed for rapid cleaning and changeover

If this level of complexity and attention to detail is required for quality in toothpaste-tube filling, just imagine what must be involved in building quality into passenger jets!

It turns out that tube-fill quality is affected by several factors, including those listed in the preceding bullet list. Problems that toothpaste manufacturers want to avoid include underfilling (mainly due to air pockets) and overfilling (resulting in tubes that "give way," to use the tube-filling industry's term; overfilling also results in giving some of the product away for free, which eats into profits). The dimensions of the inside of the tube can also play a role in tube-fill quality. For example, undersized tubes (even though they're filled with the proper amount of toothpaste) will bulge when sealed, and oversized tubes will give the appearance of being underfilled.

Understanding that quality = accuracy + consistency

Statistics are certainly involved in providing the data necessary to evaluate tube-filling equipment on each of the criteria listed in the preceding section. However, the role of statistics in the manufacturing process is most clearly emphasized by the manufacturer's criterion for quality in the tube-filling process: consistency and accuracy. The words *consistent* and *accurate* scream statistics louder than any other words in the manufacturing industry; they basically say it all.

Accuracy and consistency are monitored statistically by using control charts. A *control chart* is a specialized time chart that displays the values of the data in the order in which they were collected over time (see Chapter 4 for more information on time charts). Control charts use a line to denote where the manufacturer's specified value — or *target value* — is (this deals with the accuracy issue) and boundaries to indicate how far above or below the target the values are expected to be (this deals with the consistency part). The values being charted represent weights, volumes, or counts from individual products or, as is more often the case, they represent average weights, average volumes, or average counts from samples of products. The upper and lower boundaries of a control chart are called the *upper control limit (UCL)* and the *lower control limit (LCL)*.

For example, suppose a candy maker is filling bags of sour candies, and the target value is 50 pieces per bag, with the LCL = 45 pieces, and the UCL = 55 pieces. Suppose 8 bags of candy are sampled, their candies are counted, and the results are found to be as follows: 51 pieces, 53 pieces, 49 pieces, 51 pieces, 54 pieces, 47 pieces, 52 pieces, and 45 pieces. Figure 19-1 shows a picture of the resulting control chart for this process. This process (at least for the time being) seems to be in control.

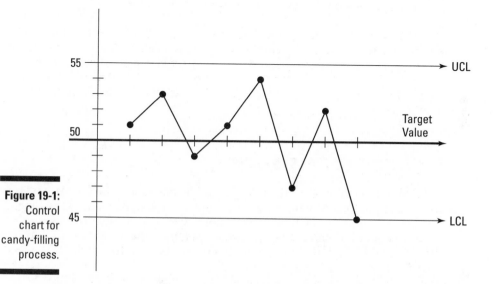

Figure 19-1:
Control
chart for
candy-filling
process.

Using control charts to monitor quality

To use statistics to monitor quality, you first have to figure out a way to define and measure accuracy and consistency. Then you have to set the target value, determine the upper and lower control limits, and collect data from the process. The data then need to be recorded on the control chart, and the process needs to be monitored to determine whether the process is in control. This last step can be a tricky decision. On the one hand, you don't want to stop the process with a false alarm (which you could be doing if you stopped it the first time a value went outside of the control limits). On the other hand, you don't want to let the process get out of control if it's beginning to produce inferior products.

Defining accuracy

What does it mean for a toothpaste tube-filling machine to be accurate? It means that the tubes weigh, in the end, what they're supposed to weigh, on average. Notice that I said "on average." I think you would agree that even the most sophisticated manufacturing process isn't perfect; some variation in any process is normal due to random fluctuation. This means that you can't expect every single tube of toothpaste to weigh exactly 6.4 ounces. However, if the tube weights start drifting lower and lower in terms of their actual weight, or if they suddenly contain a great deal of air and don't weigh what they should, consumers will notice, and the quality of the product will be compromised. Similarly, if the average weight drifts upward, the manufacturer loses money because it's giving away extra product.

Statistically speaking, product weights are *accurate* if they don't contain any bias or systematic error. (*Systematic error* leads to values that are consistently high or consistently low, compared to the expected value.) This falls right in line with the tube-filling industry's concept of accuracy. In this case the expected value (target value) is the specification set by the manufacturer, such as 6.4 ounces.

Defining consistency

What does it mean for a toothpaste tube-filling machine to be consistent? It means that the tube weights stay within the control limits most of the time. Notice I said "most of the time." Again, some variation in any process is normal due to random fluctuation. However, if the tube weights start bouncing all over the place, the quality of the product is compromised.

Statistically speaking, weights are *consistent* if their standard deviation is small (see Chapter 4). How small should the standard deviation be? It depends on the manufacturer's specifications and any limitations of the process. The operators of the tube-filling machines say that their machines are accurate to within 0.5%. This means that they expect most of the tubes labeled 6.4 ounces to weigh within 0.032 ounces of the target value (because 0.5% of 6.4 is $0.005 \times 6.4 = 0.032$ ounces). Suppose you assume that they want 95% of the tubes to be within that range. How many standard deviations are needed to cover about 95% of the values around the target?

According to the empirical rule (which you can use because you can assume the weights will have a mound-shaped distribution — see Chapter 8), 95% of the weights will be within 2 standard deviations of the target value. So, $0.032 = 2$ times the standard deviation of the weights. That means each standard deviation would be worth at most $0.032 \div 2 = 0.016$ ounces, according to the machine manufacturer's specifications. This is a conservative estimate; manufacturers' specifications may be wider than the actual control limits they set up during their quality control testing.

Expecting a normal distribution

Even though the tubes were set to be filled to 6.4 ounces, not all of the tubes are going to come out weighing exactly 6.4 ounces, of course. Some will be over, some will be under, but you expect most of the tubes to weigh close to 6.4 ounces, with an equal percentage above and below 6.4 ounces (within the control limits), if the process is in control. Beyond having a general mound shape in the middle, the weights of the toothpaste tubes should actually have a bell-shaped, or normal distribution. (See Chapter 8 for more about the normal distribution.) If the process is accurate, the mean of this distribution will be the target value, indicated by μ. And you know that the manufacturer wants the standard deviation to be no more than 0.016 ounces for consistency purposes. That means that μ = 6.4 ounces and σ = 0.016 ounces. See Figure 19-2.

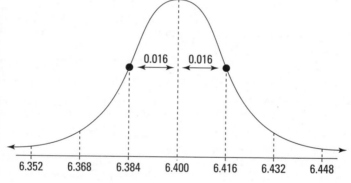

Figure 19-2:
Distribution of individual toothpaste tube weights (where the process is in control).

| 6.352 | 6.368 | 6.384 | 6.400 | 6.416 | 6.432 | 6.448 |

The standard deviation of a population is denoted by σ, and the standard deviation of a sample is denoted by s. (See Chapter 5 for more on standard deviation.) In most statistical situations, the standard deviation of a population is unknown, and you use s to estimate it from a sample. But in the case of quality control, the standard deviation for the population of products manufactured is set by the manufacturer. The standard deviation for the population is expected to be this set value, if the process is in control.

Finding the control limits

Now that the target value has been set and the expected standard deviation has been determined, the next step in statistical process control is to set the control limits for the process.

If the manufacturer is going to weigh individual tubes of toothpaste, the control limits will be set at the target value (mean) plus or minus 2 standard deviations (for 95% confidence) or the target value (mean) plus or minus 3 standard deviations (for 99% confidence). The formulas for the control limits for the individual weights are: $\mu \pm 2\sigma$ or $\mu \pm 3\sigma$, respectively.

The toothpaste manufacturers have set the mean at 6.4 and the standard deviation at 0.016. Suppose they want 95% confidence. This means that the control limits are $6.4 \pm 2(0.016) = 6.4 \pm 0.032$. The LCL is $6.4 - 0.032 = 6.368$ ounces, and the UCL is $6.4 + 0.032 = 6.432$ ounces. If the manufacturers want 99% confidence, the control limits for the weights of individual tubes are $6.4 \pm 3(0.016) = 6.4 \pm 0.048$. The LCL is $6.4 - 0.048 = 6.352$ ounces, and the UCL is $6.4 + 0.048 = 6.448$ ounces.

However, most processes are monitored by taking samples and finding the average weight of each sample, rather than looking at individual products. This means that the control limits will include the target value (mean), plus or minus 2 standard *errors* (for 95% confidence) or the target value (mean) plus or minus 3 standard *errors* (for 99% confidence). A *standard error* is the standard deviation of the sample means, and is calculated by taking the standard deviation of the weights, divided by the square root of n, where n is the sample size. (See Chapters 9 and 10 for more on the standard error.) The notation for standard error is $\frac{\sigma}{\sqrt{n}}$. The formulas for the control limits for the sample means are given by: $\mu \pm 2\frac{\sigma}{\sqrt{n}}$, or $\mu \pm 3\frac{\sigma}{\sqrt{n}}$, respectively.

The standard error will always be smaller than the standard deviation. That's because means are more consistent than individual values, because they're based on more data and, therefore, won't vary as much from one sample to the next. The larger the sample size is, the smaller the standard error is going to be (because with n in the denominator, as n increases, the value of the standard deviation divided by the square root of n decreases). See Chapter 10 for details.

If the process is monitored by weighing each tube as it comes through, that's the same as monitoring samples of size $n = 1$. So for $n = 1$, the formulas for standard deviation and standard error are the same (which they should be).

Suppose the size of each sample of toothpaste tubes is 10. Given that the standard deviation is set at 0.016, the standard error for the sample means (each of size 10) is $\frac{0.016}{\sqrt{10}} = 0.005$ ounces. The distribution of the average weights will be normal, with mean 6.4 and standard error 0.005, if the process is in control. See Figure 19-3.

Assuming that the toothpaste manufacturers are using 3 standard deviations for their acceptable level of consistency (to be conservative), the formula for the control limits is $\mu \pm 3 \dfrac{\sigma}{\sqrt{n}} = 6.4 \pm 3 \times \dfrac{0.016}{\sqrt{10}} = 6.4 \pm 3(0.005)$. The lower control limit (LCL) is $6.4 - 3(0.005) = 6.4 - 0.015 = 6.385$ ounces, and the upper control limit is $6.4 + 3(0.005) = 6.4 + 0.015 = 6.415$ ounces. The target value and control limits for this particular process are shown on the control chart in Figure 19-4.

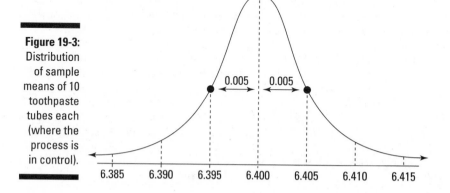

Figure 19-3:
Distribution of sample means of 10 toothpaste tubes each (where the process is in control).

A control chart uses the boundaries from the normal distribution (2 or 3 standard errors above and below the mean), but it also shows the progression of the values in time order.

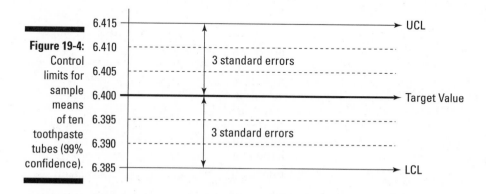

Figure 19-4:
Control limits for sample means of ten toothpaste tubes (99% confidence).

Monitoring the process

After the control limits have been set, the next step is to monitor the process. In most cases, this involves taking samples of the products at various times, finding their average weights, and marking these averages on a control chart. When it looks like the process is going off track, either in terms of accuracy or consistency, the manufacturing process is stopped, the problem is identified, and repairs or adjustments are made.

If a process is in control, you should see about 68% of the sample means falling within 1 standard error, 95% of the sample means falling within 2 standard errors and 99% of the sample means falling within 3 standard errors of the target value, according to the empirical rule (see Chapter 8). The overall sample mean should be the target value, and you should see about as many values below the target as above the target, with no particular patterns to those values.

In actual production, when the tube-filler has to be stopped, the entire production line comes to a standstill. Valuable time and production capacity are lost while maintenance and other personnel locate and correct the problem. This puts added pressure on the statistical process to always provide accurate and reliable information. Before deciding to stop the process, the workers responsible for the quality of the product want to be absolutely sure that a problem really does exist; on the other hand, if a problem has developed on the production line, they don't want to let it go on too long before making the necessary corrections. A delicate balance must be found.

So the next question is, how do you determine when a process is "out of control"? And how do you do it without having too many false alarms, costing the company time and money? Like most issues involving statistics, no single, correct, end-all and be-all answer exists (as it does for math, in most cases). Some people say that's what they love about statistics, and some say that's what they hate about statistics.

If the control limits are set at plus or minus 2 standard errors, then 95% of the means should lie within these control limits for the process to be in control. But this means that you should expect 5% of the results to lie outside these limits just by chance, and that should be okay! Here's the tricky part. You don't want to stop the process the first time a mean falls outside of the control limits; that's expected to happen 5% of the time, just by random chance (see Chapter 10 for more on this). So, in order to avoid too many false alarms, you need more than one mean to fall outside the limits to make the decision to stop the process.

You want to stop the process only if you're pretty sure something has gone awry, and one sample mean outside of the limits isn't anything out of the ordinary. Now, what if you saw 2, 3, 4, 5, or more results in a row fall outside your pre-set limits? Where's the cutoff point? Welcome to the wonderfully vague world of statistics! Here are four examples of rules that are often used to determine whether a process is out of control and should be stopped:

✔ Five sample means in a row are all either above the target or below the target (as in Figure 19-5a). Suspected cause: systematic overfilling or underfilling due to problems in the process.

✔ Six sample means in a row are either steadily increasing or steadily decreasing (as in Figure 19-5b). Suspected cause: The products coming off the line are slowly drifting farther and farther away from the intended average value, probably due to problems with one or more machines.

✔ Fourteen sample means in a row alternating above and below the target value (as in Figure 19-5c). Suspected cause: two different operators, machines, or suppliers are feeding into one system but are not in agreement.

✔ Fifteen sample means in a row are only within 1 standard error of the target (as in Figure 19-5d). Suspected cause: the process is more consistent than the specifications call for. (If this overly consistent process costs time or money, it should be loosened up. If the overly consistent process does not add time or money, finding out why the process changed — and replicating this change in the future — may be worthwhile.)

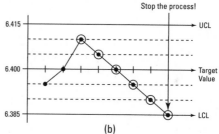

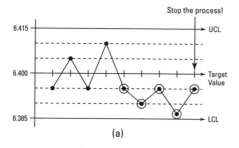

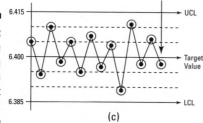

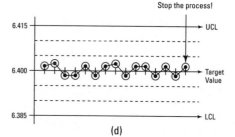

Figure 19-5: Toothpaste filling processes that are out of control.

These rules are based on probability; you stop the process when the chance of the process still being in control is very small given the data you're getting. Note that the chance of any one sample mean falling above or below the target is 50%, or 0.5. So, for the first rule listed in the preceding bullet list, the probability of getting 5 sample means in a row that are all on the same side of the target is $(0.5) \times (0.5) \times (0.5) \times (0.5) \times (0.5) = 0.03 = 3\%$. This is under the typical cutoff probability of 5% (see Chapter 14). You conclude that the process is not in control. The chance that the process is still in control, given the data, is too small.

The next time you crack open a new tube of toothpaste, think about all of the statistics that went into ensuring it was filled with quality.

Part VIII
The Part of Tens

The 5th Wave   By Rich Tennant

AT THE STATISTICIANS' DATING BAR

Whoa! Look at the pocket protectors on this one!

In this part . . .

*W*here would a statistics book be without some statistics of its own? This part contains ten criteria for a good survey and ten common statistical mistakes.

This part gives you a quick, concise reference that you can use to help critique or design a survey and detect common statistical abuses.

Chapter 20

Ten Criteria for a Good Survey

In This Chapter
▶ Critically evaluating a survey
▶ Planning a good survey

Surveys are all around you: I guarantee that at some point in your life, you'll be asked to complete a survey. This means that you're also inundated with the results of those surveys, and before you consume the information, you need to evaluate whether a survey was properly designed and implemented — in other words, don't assume the survey is okay until you check it out (see Chapter 16 for the lowdown on surveys). The two important goals for a survey are to be *accurate* (that is, based on enough data so the results wouldn't change much if another sample were taken) and to have a minimum amount of *bias* (systematic overestimation or underestimation of the true result, like the bathroom scale that is always five pounds too high!). In this chapter, you find ten criteria that you can use to evaluate or plan a survey.

The Target Population Is Well Defined

The *target population* is the entire group of individuals that you're interested in studying. For example, suppose you want to know what the people in Great Britain think of reality TV. The target population in this case would be all the residents of Great Britain.

Note that sometimes, the target population needs a bit of refinement for clarity. For example, what age groups do you want to include in your target population? For the reality TV example, you probably don't want to include children under a certain age, say 12. So your target population is actually all residents of Great Britain aged 12 and over.

Many researchers don't do a good job of defining their target populations clearly. For example, if the American Egg Board wants to say "Eggs are good for you!" it needs to specify who the "you" is. For example, is the Egg Board prepared to say that eggs are good for people who have high cholesterol?

(One of the studies the group cited was based only on young people who were healthy and eating low-fat diets — is that who they mean by "you"?)

If the target population isn't well defined, the survey results are likely biased. This is because the sample that's actually studied may contain people outside the intended population, or the survey may exclude people who should have been included.

The Sample Matches the Target Population

When you're conducting a survey, you typically can't ask every single member of the target population to provide the information you're looking for. You usually don't have the time or money to do that. The best you can do is select a *sample* (a subset of individuals from the population) and get the information from them. Because this sample of individuals is your only link to the entire target population, you want that sample to be really good.

A good sample represents the target population. The sample doesn't systematically favor certain groups within the target population, and it doesn't systematically exclude certain people, either. This sounds easy enough, right? All you need to do is get a list of all the individuals in the target population (this is called a *sampling frame*) and select a sample of people from it. How difficult can that be?

Pretty difficult. Suppose your target population is all registered voters in the United States who are likely to vote in the next presidential election. Getting a list of these individuals isn't easy. You can look at voter registration lists, but you don't know which people are likely to vote in the next election. You could check out those who voted in the last election, but many of those folks moved or died, and you're not including those who turned 18 since the last election. Suddenly, the situation gets a bit complicated. Welcome to the world of surveys!

One potential solution to this problem is to obtain updated voter registration lists, take a sample of individuals from those lists, and ask them whether they plan to vote in the upcoming election. When someone doesn't, stop asking questions and don't count that person in your survey. For those who do plan to vote, ask who they plan to vote for, and include those answers in your survey results.

A good survey has an updated and comprehensive sampling frame that lists all the members of the target population, if possible. If such a list isn't possible, some mechanism is needed that gives everyone in the population an equal opportunity to be chosen to participate in the survey. For example, if a house-to-house survey of a city is needed, an updated map including all houses in that city should be used as the sampling frame.

The Sample Is Randomly Selected

An important feature of a good survey is that the sample is randomly selected from the target population. *Randomly* means that every member of the target population has an equal chance of being included in the sample. In other words, the process you use for selecting your sample can't be biased.

Suppose you have a herd of 1,000 steers, and you need to take a random sample of 50 of them to test for a disease. Taking the first 50 steers that come up to you in the field wouldn't fit the definition of a random sample. The steers that are able to come up to you may be less likely to have any kind of disease, or they may be the older, more friendly ones, who actually may be more susceptible to disease. Either way, bias is introduced in the survey. How do you take a random sample of steers? The animals are likely tagged with ID numbers, so you get a list of all the ID numbers, take a random sample of those, and locate those animals. Or, if the animals sleep in cages or stalls, number those and take a random sample of cage numbers. Sometimes being a statistician means being very inventive about how you take a truly random sample!

For surveys involving people, reputable polling organizations such as The Gallup Organization use a random digit dialing procedure and telephone the members of their sample. This excludes people without phones, of course, so this survey has a bit of bias. In this case, though, most people do have phones (over 95%, according to The Gallup Organization), so the bias against people who don't have phones is not a big problem.

A good survey contains a random sample of individuals from the target population. Be sure to find out how the sample was selected, if that process isn't described.

The Sample Size Is Large Enough

You've heard the saying, "Less is more"? With surveys, the saying is, "Less good information is better than more bad information, but more good information is better." (Not really catchy, is it?)

Here's the basic idea. If you have a large sample size, and the sample is representative of the target population (meaning randomly selected), you can count on that information to be pretty accurate. How accurate depends on the sample size, but the bigger the sample size, the more accurate the information will be.

A quick and dirty formula to calculate the accuracy of a survey is to divide by the square root of the sample size. For example, a survey of 1,000 (randomly selected) people is accurate to within $\frac{1}{\sqrt{1,000}}$, which is 0.032 or 3.2%. This percentage is called the *margin of error*.

Beware of surveys that have a large sample size that *is not* randomly selected. Internet surveys are the biggest culprit here. A company can say that 50,000 people logged on to its Web site to answer a survey, which means the survey results have a lot of information. But that information is biased, because it doesn't represent the opinions of anyone except those who chose to participate in the survey; that is, they had access to the Internet, went to the Web site, and chose to complete the survey. In this case, less would have been more: The company should have sampled fewer people but done so randomly.

Good Follow-Up Minimizes Non-Response

After the sample size has been chosen and the sample of individuals has been randomly selected from the target population, you have to get the information you need from the people in the sample. If you've ever thrown away a survey or refused to answer a few questions over the phone, you know that getting people to participate in a survey isn't easy. If a researcher wants to minimizes bias, the best way to handle non-responses is to "hound" the people in the sample: Follow up one, two, or even three times, offering dollar bills, coupons, self-addressed stamped return envelopes, chances to win prizes, and so on. Note that offering more than a small token of incentive and appreciation for participating would create bias, as well, because then people who really need the money are more likely to respond than those who don't.

Consider what motivates you to fill out a survey. If the incentive provided by the researcher doesn't get you, maybe the subject matter peaks your interest. Unfortunately, this is where bias comes in. If only those folks who feel very strongly respond to a survey, only their opinions will count; because the other people who don't really care about the issue don't respond, each "I don't care" vote doesn't count. And when people do care but don't take the time to complete the survey, those votes doesn't count, either.

For example, suppose 1,000 people are given a survey on whether to change the park rules to allow dogs. Who will respond? Most likely, the respondents will be those who feel strongly for allowing dogs and who feel strongly against dogs in the park. Suppose 100 of each are the only respondents and the other 800 surveys aren't returned. This means that 800 opinions aren't counted. If none of those 800 people really cares about the issue either way and if you could count their opinions, the results would be 800 ÷ 1000 or 80% saying "no opinion," with 10% (100 ÷ 1,000) saying they support the issue and 10% (100 ÷ 1,000) against the issue. Without those votes of the 800 non-respondents, however, researchers can report, "Of the people who responded, 50% were for the issue, and 50% were against it." This gives the impression of a very different (and biased!) result.

The *response rate* of a survey is a percentage found by taking the number of respondents divided by the total sample size and multiplying by 100%. A good response rate is anything over 70%. However, most response rates fall well short of that, unless the survey is done by a very reputable organization, such as The Gallup Organization. Look for the response rate when examining survey results. If the response rate is too low (much less than 70%) the results may be biased and should be ignored.

Selecting a smaller initial sample and following up aggressively is better than selecting a bigger sample that ends up with a low response rate. Aggressive follow-up reduces bias.

The next time you're asked to participate in a good survey (according to the criteria listed in this chapter), consider responding. You'll be doing your part to rid the world of bias!

The Type of Survey Used Is Appropriate

Surveys come in many types: mail surveys, telephone surveys, Internet surveys, house-to-house interviews, and man-on-the-street surveys (in which someone comes up to you with a clipboard and asks, "Do you have a few minutes to participate in a survey?"). One very important yet sometimes overlooked criterion of a good survey is whether the type of survey being used is appropriate for the situation.

For example, if the target population is the population of people who are visually impaired, sending them a survey in the mail that has a tiny font isn't a good idea (yes, this has happened!). If you want to conduct a survey of victims of domestic violence, interviewing them in their homes isn't appropriate.

Suppose your target population is the homeless people in your city. How do you contact them? They don't have addresses or phones, so neither type of survey is appropriate. (This is a very difficult problem, and one that the government wrestles with when census time comes around.) You can physically go and talk to people one on one, wherever they are located, but to find out where they're located is no easy task. Asking local shelters, churches, or other groups who help the homeless may give you some good leads to start the search.

When looking at the results of a survey, be sure to find out what type of survey was used and reflect on whether this type of survey was appropriate.

The Questions Are Well Worded

The way that a question is worded in a survey can affect the results. For example, while President Bill Clinton was in office and the Monica Lewinsky scandal broke, a CNN/Gallup Poll conducted August 21–23, 1998, asked respondents to judge Clinton's favorability, and about 60% gave him a positive result. (The sample size for this survey was 1,317, and the margin of error was reported to be plus or minus 3 percentage points.) When CNN/Gallup reworded the question to ask respondents to judge Clinton's favorability "as a person," the results changed: Only about 40% gave him a positive rating.

The next night, CNN/Gallup conducted another survey on the same topic. Here are some of the questions and responses:

- ✔ Do you approve or disapprove of the way President Clinton is handling his job? (Sixty-two percent approved and 35% disapproved.)

- ✔ Do you have a favorable or unfavorable overall opinion of Clinton? (Fifty-four percent said favorable, 43% said unfavorable.)

- ✔ All things considered, are you glad Clinton is president? (Fifty-six percent said yes, 42% said no.)

- ✔ If you could vote again for the 1996 candidates for president, who would you vote for? (Forty-six percent said they would vote for Bill Clinton, 34% for Bob Dole, 13% for Ross Perot.)

These questions are all getting at the same issue: how people felt about President Clinton during the time of the Monica Lewinsky scandal. And while these questions are similar, all are worded slightly differently, and you can see how different the results are. So question wording does matter.

Probably the biggest problem with question wording though is the use of *leading questions.* In other words, questions that are worded in such a way that you know how the researcher wants you to answer. This leads to biased results that give too much weight to a certain answer because of the wording of the question, not because people actually have that opinion.

Many surveys contain leading questions (either unintentionally or by design), to get you to say what the pollster wants you to say. Here are a couple of examples of leading questions similar to those I've seen in print:

- ✔ Which position most agrees with yours? Democrats favor responsible, realistic fiscal planning to balance the budget in a reasonable period of time, while still meeting their responsibilities to the most vulnerable Americans. Republicans propose to enforce a mandatory balanced budget while allowing for severe cuts in education and health care.

- ✔ Do you agree that the president should have the power of a line-item veto to eliminate waste?

- ✔ Do you think that Congress and White House officials should set a good example by eliminating the perks and special privileges they currently receive at each taxpayer's expense?

When you see the results of a survey that's important to you, ask for a copy of the questions that were asked and analyze them to ensure that they were neutral and minimized bias.

The Survey Is Properly Timed

The timing of a survey is everything. Current events shape people's opinions, and while some pollsters try to determine how people really feel, others take advantage of these situations, especially the negative ones. For example, when shootings have occurred in schools, the issue of gun control has often been raised in surveys. Of course, during the time immediately following such a tragedy, more people are in favor of gun control than before; in other words, the results spike upward. After a period of time, however, opinions go back to what they were before; meanwhile the pollsters project the results of the survey as if the public feels that way all the time.

Timing of any survey, regardless of the subject matter, can still cause bias. For example, suppose your target population is people who work full time. If you conduct a telephone survey by calling their homes between the hours of 9 a.m. and 5 p.m., you're going to have a lot of bias in your results, because those are the hours that the majority of fulltime workers are at work!

Check the date when a survey was conducted and see whether you can determine any relevant events that may have temporarily influenced the results. Also look at what time of the day or night the survey was conducted: Was it during a time that is most convenient for the target population to respond?

The Survey Personnel Are Well Trained

The people who actually carry out surveys have tough jobs. They have to deal with hang ups, take-us-off-your-list responses, and answering machines. And then, after they do get a live respondent at the other end of the phone or face to face, their job becomes even harder. That's when they have to collect data in an accurate and unbiased way.

Here are some problems that can come up during the survey process:

- ✔ The respondent doesn't understand the question and needs more information. How much do you tell that person, while still remaining neutral?

- ✔ Information gets miscoded. For example, I tell the survey person I'm 40 years old, but he accidentally writes down 60.

- ✔ The person carrying out the survey has to make a judgment call. For example, suppose the question asks how many people are living in the house, and the respondent asks, "Do I count my cousin Bob who is just staying with us while he looks for a job?" A decision needs to be made.

- ✔ Respondents may give misinformation. For example, some people hate surveys so much that they go beyond refusing to do them and actually complete them, but give all misleading answers. For example, when asked how old she is, a woman may say 101 years old.

How do the survey personnel handle these and a host of other challenges that occur during the survey process? The key is to be clear and consistent about every possible scenario that may come up, discuss how they should be handled, and have this discussion well before participants are ever contacted. That means the personnel need to be well trained.

You can also avoid problems by running a *pilot study* (a practice run with a only a few respondents) that's recorded, so that personnel can practice and be evaluated on how accurate and consistent they are in collecting their data. In this way, researchers can anticipate problems before the survey process starts and put policies in place for how they are to be handled.

Surveys should avoid unclear, ambiguous, or leading questions. A pilot study can also screen potentially difficult questions by gauging how the small group responds and what questions they ask. And to avoid miscoding the information (that is, typing the information incorrectly), the survey needs to have all possible answers clearly marked. For example: If strongly disagree is to be coded with a 1, and strongly agree is to be coded with a 5 (and not the other way around) the survey needs to clearly specify that. Taping the interview for a crosscheck later would also help.

Be sure you've decided how to handle *prank respondents,* people who are merely joking around. One suggestion is to flag that person's response as possibly unusable, and then call the number back later and try again.

The Survey Answers the Original Question

Suppose a researcher starts out with a statement like, "I want to find out about shoppers' buying habits." That sounds good, right? But then you look at the survey, and the questions are all about how people feel about shopping ("What do you like best/least about shopping?" or "On a scale of 1–10, how much do you enjoy shopping?"). The questions don't ask about buying behavior. While attitudes toward shopping do influence shoppers' buying habits, the real measure of buying habits is how shoppers behave: what they're buying, how much they spend, where they're shopping, with whom they shop, when they shop, how often they shop, and so on. Sometimes, researchers don't realize that they have missed the boat until after their survey results are in. After the fact (and too late for them to do anything), they see that they can't answer their research questions with the data they collected, which is not good!

Before participating in a survey, ask what the researcher is trying to find out — what the purpose of the survey is. Then as you read or listen to the questions, if they appear to have an agenda or to be leading you in a different direction, my advice is to stop your participation and explain why, either in writing or in person.

Before designing any survey, first write down the goals of the survey. What do you want to know? Then design the questions to meet those goals. That way, you're sure to get your questions answered.

Chapter 21

Ten Common Statistical Mistakes

....

In This Chapter

▶ Recognizing common statistical mistakes made by researchers and the media

▶ Avoiding mistakes when doing your statistics

....

*T*his book is about not only understanding the statistics that you come across in the media and in your workplace, but digging deeper to examine whether those statistics are correct, reasonable, and fair. You have to be vigilant — and a bit skeptical — to deal with today's information explosion, because many of the statistics you come across are wrong or misleading, either by error or by design. If you don't critique the information you're consuming, who will? In this chapter, I outline some common statistical mistakes made by researchers and by the media, and I share ways to recognize and avoid those mistakes.

Misleading Graphs

Many graphs and charts contain misinformation, mislabeled information, or misleading information, or they simply lack important information that the reader needs to make critical decisions about what is being presented. Figure 21-1 shows examples of four types of data displays: pie charts, bar graphs, time charts, and histograms. (Note that a *histogram* is basically a bar graph for numerical data.) For each type, I outline some of the most common ways that you can be misled. (For more information on charts and graphs, including misleading charts and graphs, see Chapter 4.)

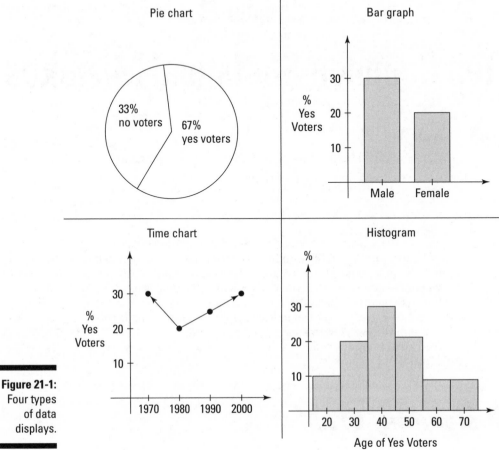

Figure 21-1:
Four types
of data
displays.

Pie charts

Pie charts are exactly what they sound like: charts that are in the shape of a circle (or pie) and divided into slices that represent the percentage of individuals that fall into each group (according to some categorical variable, such as gender, political party, or employment status).

Here's how to sink your teeth into a pie chart and test it for quality:

- ✔ Check to be sure the percentages add up to 100%, or close to it (any round-off error should be small).

- ✔ Beware of slices of the pie called "other" that are larger than the rest of the slices. This means the pie chart is too vague.

✔ Watch for distortions that come with the three-dimensional-looking pie charts, in which the slice closest to you looks larger than it really is because of the angle at which it's presented.

✔ Look for a reported total number of individuals who make up the pie chart, so you can determine how big the pie was before it was divided up into the slices that you're looking at. If the size of the data set (the number of respondents) is too small, the information isn't reliable.

Bar graphs

A bar graph is similar to a pie chart, except that instead of being in the shape of a circle that's divided up into slices, a bar graph represents each group as a bar, and the height of the bar represents the number or percentage of individuals in that group.

When examining a bar graph:

✔ Consider the units being represented by the height of the bars and what the results mean in terms of those units. For example, total number of crimes verses the crime rate, which is measured by total number of crimes *per capita* (that is, per person).

✔ Evaluate the appropriateness of the *scale,* or amount of space between units expressing the number in each group of the bar graph. For example, number of customer complaints for men versus women could be expressed by units of 1, 5, 10, or 100. Small scales (for example, going from 1 to 500 by 10s) make differences look bigger; large scales (for example, going from 1 to 500 by 100s) make differences look smaller.

Time charts

A time chart shows how some measurable quantity changes over time (for example, stock prices, average household income, and average temperature).

Here are some issues to watch for with time charts:

✔ Watch the scale on the vertical (quantity) axis as well as the horizontal (timeline) axis; results can be made to look more or less dramatic than they actually are by simply changing the scale.

✔ Take into account the units being portrayed by the chart and be sure they are equitable for comparison over time; for example, are dollar amounts being adjusted for inflation?

✔ Beware of people trying to explain why a trend is occurring without additional statistics to back themselves up. A time chart generally shows *what* is happening. *Why* it's happening is another story!

✔ Watch for situations in which the time axis isn't marked with equally spaced jumps. This often happens when data are missing. For example, the time axis may have equal spacing between 1971, 1972, 1975, 1976, 1978, when it should actually show empty spaces for the years in which no data are available.

Histograms

A histogram is a graph that breaks the sample into groups according to a numerical variable (such as age, height, weight, or income) and shows either the number of individuals (frequency) or the percentage of individuals (relative frequency) that fall into each group. For example, perhaps 20% of a sample was 20 years old or less; 30% was between 20 and 40 years old; 45% was between 40 and 60 years old; and 5% was over 60 years old (age being the numerical variable here).

Some items to watch for regarding histograms include the following:

✔ Watch the scale used for the vertical (frequency/relative frequency) axis, looking especially for results that are exaggerated or played down through the use of inappropriate scales.

✔ Check out the units on the vertical axis, whether they're reporting frequencies or relative frequencies; take that information into account when examining the information.

✔ Look at the scale used for the groupings of the numerical variable on the horizontal axis. If the groups are based on small intervals (for example, 0–2, 2–4, and so on), the data may look overly volatile. If the groups are based on large intervals (for example, 0–100, 100–200, and so on), the data may give a smoother appearance than is realistic.

Biased Data

Bias in statistics is the result of a systematic error that either overestimates or underestimates the true value. For example, if I use a ruler to measure plants and that ruler is ½-inch short, all of my results are biased; they're systematically lower than their true values.

Here are some of the most common sources of biased data:

✔ **Measurement instruments are systematically off.** For example, a police officer's radar gun says you were going 76 miles per hour but you *know* you were only going 72 miles per hour. Or a scale that always adds 5 pounds to your weight.

✔ **Participants are influenced by the data collection process.** For example, a survey question that asks, "Have you *ever* disagreed with the government?" will overestimate the percentage of people who are unhappy with the government.

✔ **Sample of individuals doesn't represent the population of interest.** For example, examining study habits by visiting only the campus library puts more emphasis on those who study the most. (See more on this in the "Non-Random Samples" section later in this chapter.)

✔ **Researchers aren't objective.** For example, suppose one group of patients is given a sugar pill, and the other group is given a real drug. The researchers can't know who got what treatment, because if they do, they may inadvertently project results onto the patients (such as saying, "You're feeling better, aren't you?") or pay more attention to those on the drug.

To spot biased data, examine how the data were collected. Ask questions about the selection of the participants, how the study was conducted, what questions were used, what *treatments* (medications, procedures, therapy, and so on) were given (if any), and who knew about them, what measurement instruments were used and how they were calibrated, and so on. Look for systematic errors or favoritism, and if you see too much of it, ignore the results.

No Margin of Error

The word "error" has a somewhat negative connotation, as if an error is something that is always avoidable. In statistics, that's not always the case. For example, a certain amount of what statisticians call *sampling error* will always occur whenever someone tries to estimate a population value using anything other than the entire population. Just the act of selecting a sample from the population means you leave out certain individuals, and that means you're not going to get the precise, exact population value. No worries, though. Remember that statistics means never having to say you're certain — you have to only get close. And if the sample is large and is randomly selected, the sampling error will be small.

To evaluate a statistical result, you need a measure of its accuracy — typically through the margin of error. The *margin of error* tells you how much the researcher expects his or her results to vary from sample to sample. (For more information on margin of error, see Chapter 10.) When a researcher or the media fail to report the margin of error, you're left to wonder about the accuracy of the results, or worse, you just assume that everything is fine, when in many cases, it's not. Survey results shown on TV used to only rarely include a margin of error, but now, you do often see that information reported. Still, many newspapers, magazines, and Internet surveys fail to report a margin of error, or they report a margin of error that is meaningless because the data are biased (see Chapter 10).

When looking at statistical results in which a number is being estimated (for example, the percentage of people who think the president is doing a good job) always check for the margin of error. If it's not included, ask for it! (Or if given enough other pertinent information, you can calculate the margin of error yourself using the formulas in Chapter 10.)

Non-Random Samples

From survey data to results of medical studies, most statistics are based on data collected from samples of individuals, rather than on entire populations, because of cost and time considerations. Plus, you don't need a very big sample to be amazingly accurate, *if* the sample being studied is representative of the population. For example, a well-designed and well-conducted survey of 2,500 people has a margin of error of roughly plus or minus only 2% (see Chapter 10). For an experiment with a treatment group and a control group, statisticians would like to see at least 30 people in each group in order to have accurate data.

How can you ensure that the sample represents the population? The best way is to *randomly* select the individuals from the population. A random sample is a subset of the population selected in such a way that each member of the population has an equal chance of being selected (like drawing names out of a hat). No systematic favoritism or exclusion is involved in a random sample.

Many surveys and studies aren't based on random samples of individuals. For example, medical studies are often performed with volunteers that, obviously, volunteer — they're not randomly selected. It wouldn't be practical to phone people and say, "You were chosen at random to participate in our sleeping study. You'll need to come down to our lab and stay there for four nights." In this situation, the best you can do is study the volunteers and see how well they represent the population, and then report those results. You can also ask for certain types of volunteers.

Polls and surveys also need to be based on randomly selected individuals, and doing so is much easier than with medical studies. However, many surveys aren't based on random samples. For example, TV polls asking viewers to "call us with your opinion" aren't based on random samples. These surveys don't give the entire population an equal chance to be chosen (in fact, in these examples, the people choose themselves).

Before making any decisions about statistical results from a survey or a study, look to see how the sample of individuals was selected. If the sample wasn't selected randomly, take the results with a grain of salt.

Missing Sample Sizes

The quantity of information is always important in terms of assessing how accurate a statistic will be. The more information that goes into a statistic, the more accurate that statistic will be — as long as that information isn't biased, of course (see the "Biased Data" section earlier in this chapter). The consumer of the statistical information needs to assess the accuracy of the information, and for that, you need to look at how the information was collected (see Chapter 16 regarding surveys and Chapter 17 regarding experiments), and how much information was collected (that is, you have to know the sample size).

Many charts and graphs that appear in the media don't include a sample size. You also find that many headlines are "not exactly" what they seem to be, when the details of an article reveal either a small sample size (reducing reliability in the results) or in some cases, no information at all about the sample size. (For example, you've probably seen the chewing gum ad that says, "Four out of five dentists surveyed recommend [this gum] for their patients who chew gum." What if they really did ask only five dentists?)

Always look for the sample size before making decisions about statistical information. The smaller the sample size, the less reliable the information. If the sample size is missing from the article, get a copy of the full report of the study, contact the researcher, or contact the journalist who wrote the article.

Misinterpreted Correlations

The statistical definition of *correlation* is the strength and direction of the linear relationship between two numerical variables. In other words, the amount by which one numerical variable (for example, weight) is expected to increase or decrease if another numerical variable (for example, height) is allowed to increase/decrease. Correlation is one of the most misunderstood and misused statistical terms used by researchers, the media, and the general public. Three important points about correlation are as follows:

- ✓ **A correlation can't apply to two *categorical variables,* such as political party and gender. Instead, correlation applies only to two *numerical variables,* such as height and weight.** So, if you hear someone say, "It appears that the voting pattern is correlated with gender," you know that's incorrect. Voting pattern and gender may be associated, but they can't be correlated, according to the statistical definition of correlation.

- ✓ **A correlation measures the strength and direction of the *linear* relationship between two numerical variables.** In other words, if you collect data on two numerical variables (such as height and weight) and plot all of the points on a graph, when a correlation exists, you should be able to draw a straight line through those points (uphill or downhill), and the line should fit pretty well. If a line doesn't fit, the variables aren't correlated. However, that doesn't necessarily mean the variables aren't related. They may have some other type of relationship; they just don't have a linear relationship. For example, bacteria multiply at an *exponential* rate over time (their numbers explode, doubling faster and faster), not at a *linear* rate (which would be a steady increase over time).

- ✓ **Correlation doesn't automatically mean cause and effect.** For example, suppose someone reports that more people who drink diet soda have brain tumors than people who don't. If you're a diet soda drinker, don't panic just yet. This may be a freak of nature, and someone happened to notice it. At most, it means more research needs to be done (beyond observation) in order to show diet soda *causes* brain tumors.

Confounding Variables

A *confounding variable* is a variable that wasn't included in a study that may influence the results of the study, creating a confusing (that is, confounding) effect. For example, suppose a researcher tries to say that eating seaweed helps you live longer, and when you examine the study further, you find out that it was based on a sample of people who regularly eat seaweed in their diets and are over the age of 100. Suppose you then read interviews of these people and discover some of their other secrets to long life (besides eating seaweed): They ate very healthy foods, slept an average of 8 hours a day, drank a lot of water, and exercised every day. So did the seaweed cause them to live longer? Maybe, but you can't tell, because the confounding variables (exercise, water consumption, diet, and sleeping patterns) could also have caused longer life.

A common error in research studies is to fail to control for confounding variables, leaving the results open to scrutiny (and you want to be among those doing the scrutinizing)! The best way to control for confounding variables is to do a well-designed *experiment,* which involves setting up two groups that are alike in as many ways as possible, except that one group (called the *treatment group*) takes a treatment, and the other takes a *placebo* (a fake treatment; this group is called the *control group*). You then compare the results from the two groups, attributing any significant differences to the treatment (and to nothing else, in an ideal world).

The seaweed study wasn't a designed experiment; it was an *observational study.* In observational studies, no control for any variables exists; people are merely observed, and information is recorded.

Whenever you're looking at the results of a study that's claiming to show a cause-and-effect relationship or significant differences between groups, check to see whether the study was a designed experiment and whether confounding variables were controlled for. If doing an experiment is unethical (for example, showing smoking causes lung cancer by forcing half of the subjects in the experiment to smoke 10 packs a day for 20 years, while the other half of the subjects smoke nothing) you must rely on mounting evidence from many observational studies that cover many different situations, all leading to the same result.

Observational studies are great for surveys and polls, but not for showing cause-and-effect relationships, because they don't control for confounding variables. A well-designed experiment provides much stronger evidence.

Botched Numbers

Just because a statistic appears in the media doesn't mean it's correct. In fact, errors appear all the time (by error or by design), so stay on the lookout for them. Here are some tips for spotting botched numbers:

- ✓ **Make sure everything adds up to what it's reported to.** With pie charts, be sure all the percentages add up to 100%.

- ✓ **Double check even the most basic of calculations.** For example, a chart says 83% of Americans are in favor of an issue, but the report says 7 out of every 8 Americans are in favor of the issue. Are these the same? (No, 7 divided by 8 is 87.5%; 5 out of 6 is about 83%.)

- ✓ **Look for the response rate of a survey; don't just be happy with the number of participants.** (The *response rate* is the number of people who responded divided by the total number of people surveyed times 100%.) If the response rate is much lower than 70%, the results could be biased, because you don't know what the non-respondents would have said.

✔ **Question the type of statistic used, to determine whether it's appropriate.** For example, the number of crimes went up, but so did the population size. The researchers should have reported the *crime rate* (number of crimes per capita), instead.

Statistics are based on formulas and calculations that don't know any better — the people plugging in the numbers should know better, though, and they either don't know better or they don't want you to catch on. You, as a consumer of information (also known as a certified skeptic), must be the one to take action. The best policy is to ask questions.

Selectively Reporting Results

Another bad scenario is when a researcher reports his one *statistically significant result* (a result that was very unlikely to have occurred simply by chance), but leaves out the part where he actually conducted hundreds of other, unreported tests, all of which were *not* significant. If you had known about all of the other tests, you may have wondered whether this one that was significant was meaningful, or if, indeed, it was simply due to chance because of the large number of tests performed. This is what statisticians like to call *data snooping* or *data fishing*.

How do you protect yourself against misleading results due to data fishing? Find out more details about the study: how many tests were done, how many results weren't significant, and what was found to be significant. In other words, get the whole story if you can, so that you can put the significant results into perspective.

To spot fudged numbers and errors of omission, the best policy is to remember that if something looks too good to be true, it probably is. Don't just go on the first result that you hear, especially if it makes big news. Wait to see whether others can verify and replicate the result.

The Almighty Anecdote

Ah, the anecdote — one of the strongest influences on public opinion and behavior ever created. And one of the least statistical. An *anecdote* is a story based on a single person's experience or situation. For example:

- ✔ The waitress who won the lottery
- ✔ The cat that learned how to ride a bicycle
- ✔ The woman who lost a hundred pounds in two days on the miracle potato diet
- ✔ The celebrity who claims to have used an over-the-counter hair color for which she is a spokesperson (yeah, right)

Anecdotes make great news; the more sensational the better. But sensational stories are outliers from the norm of life. They don't happen to most people.

You may think you're out of reach of the influence of anecdotes. But what about those times when you let one person's experience influence you? Your neighbor loves his Internet Service Provider, so you try it, too. Your friend had a bad experience with a certain brand of car, so you don't bother to test-drive it. Your dad knows somebody who died in a car crash because they were trapped in the car by their seat belt, so he decides never to wear his.

While some decisions are okay to make based on anecdotes, some of the more important decisions you make should be based on real statistics and real data that come from well-designed studies and careful research.

 An anecdote is a data set with a sample size of only one. You have no information with which to compare the story, no statistics to analyze, no possible explanations or information to go on — just a single story. Don't let anecdotes have much influence over you. Instead, rely on scientific studies and statistical information based on large random samples of individuals who represent their target populations (not just a single situation).

 The best thing to do when someone tries to persuade you by telling you an anecdote is to respond by saying, "Show me the data!"

Appendix

Sources

• •

*T*his appendix contains the sources that I use in my examples throughout this book. Because you're a budding statistical detective, you may want to follow up on some of these to get more information that leads you to more informed decisions.

Chapter 1

All newspaper articles mentioned were taken from *The Cincinnati Enquirer* and *The Columbus Dispatch,* January 26, 2003.

Chapter 2

Microwaving leftovers survey: *USA Today,* September 6, 2001.

Trident Gum Web site: www.tridentgum.com/consumer/html/c0000.html.

Number of crimes in the U.S. from 1990–1998, taken from the FBI Uniform Crime Reports: www.fbi.gov/ucr/ucr.htm.

Kansas Lottery Web site: www.kslottery.com.

Good bedside manner can fend off malpractice suits: *USA Today,* February 19, 1997.

Ross Perot's survey: *TV Guide,* March, 21, 1993; for more information, contact United We Stand America at: www.uwsa.com.

Journal of the American Medical Association: jama.ama-assn.org.

The New England Journal of Medicine: http:content.nejm.org.

The Lancet: www.thelancet.com.

British Medical Journal: http://bmj.com.

The Gallup Organization: www.gallup.com.

Chapter 3

The Gallup Organization: www.gallup.com.

U.S. Census Bureau: www.census.gov.

Zinc for colds; pillow position and sleep; shoe heel height and feet comfort: these studies are referenced in "Healthy Habits — that Aren't," *Woman's Day,* February 11, 2003.

Cricket chirps and temperature: "Cricket thermometers," *Field & Stream,* July 1993, Vol. 98, Issue 3, p. 21; for data, see: *The Songs of the Insects* (1949), by George W. Pierce, Harvard University Press, pp. 12–21.

Crimes and police, U.S. Dept. of Justice: www.ojp.usdoj.gov/bjs/lawenf.htm.

Ice cream and murders (New York City): a good article to start investigating this subject and related issues is Spellman, B. A., & Mandel, D. R. (2003). For more on the psychology of causal reasoning, see Nadel, L. (Ed.) *Encyclopedia of Cognitive Science* (Vol. 1, pp. 461–466).

Chapter 4

Consumer Expenditure Survey, Bureau of Labor Statistics: www.bls.gov/cex.

Lotteries: (Ohio) www.ohiolottery.com; (Florida) www.flalottery.com/lottery/edu/edu.html; (Michigan) www.michigan.gov/lottery; (New York) www.nylottery.org.

Tax dollar pizza: www.irs.gov/app/cgi-bin/slices.cgi.

Population/racial/workforce trends, U.S. Department of Labor, Herman Report: "Futurework: Trends and Challenges for Work in the 21st Century": www.dol.gov/asp/programs/history/herman/reports/futurework.

Transportation expenses, Bureau of Transportation Statistics: www.bts.gov/publications/transportation_in_the_united_states/pdf/teconomy.pdf.

Birth statistics, Colorado Department of Public Health and Environment: www.cdphe.state.co.us/../cohid/birthdata.html.

Internal Revenue Service: www.irs.gov.

Occupation employment and wage estimates, Bureau of Labor Statistics: www.wa.gov/esd/lmea/occdata/oeswage/Page2067.htm.

Population estimates by state, U.S. Census Bureau: http://eire.census.gov/popest/data/states/tables/ST-EST2002-01.php.

Chapter 5

U.S. population data, U.S. Census Bureau: www.census.gov/population/www/documentation/twps0038.pdf.

NBA player salaries: www.hoopsworld.com/article_21.shtml.

Making friends in cyberspace: Parks and Floyd (1996), *Journal of Communication* 46(1), 80–97.

Household income, U.S. Census Bureau, *Money Income in the United States,* 2001, pp. 26–27.

Chapter 6

American Community Survey, Columbus OH 2001: www.census.gov/acs/www/Products/Profiles/Single/2001/ACS/Narrative/155/NP15500US3918000049.htm.

U.S. Census Bureau: www.census.gov.

Chapter 7

Connecticut Powerball Lottery: www.ctlottery.org/powerball.htm.

Chapter 8

No sources were referenced in this chapter.

Chapter 9

Consumer expenditure survey standard errors: `www.bls.gov/cex/2001/stnderror/age.pdf`.

Bureau of Labor Statistics: `www.bls.gov`.

ACT scores and standard deviations (2002): `www.act.org/news/data/02/pdf/t6-7-8.pdf`.

ACT tables 2002 `www.act.org/news/data/02/pdf/data.pdf`.

Chapter 10

The Gallup Organization: `www.gallup.com`.

Chapter 11

Median household income confidence intervals: `www.census.gov/hhes/income/income01/statemhi.html`; **full report:** `www.census.gov/prod/2002pubs/p60-218.pdf`.

Chapter 12

Prevalence of teen drug use: "Monitoring the Future," sponsored by the National Institute of Drug Abuse and conducted by the University of Michigan. `http://monitoringthefuture.org/data/2002data-drugs`.

Chapter 13

No sources were referenced in this chapter.

Chapter 14

Varicose veins: *Woman's Day,* Feb 11, 2003 p. 28.

Baby sleep: *Woman's Day,* Feb 11, 2003, p. 120 "And So, to Bed" by Loraine Stern.

Prevalence of teen drug use: "Monitoring the Future," sponsored by the National Institute of Drug Abuse and conducted by the University of Michigan. http://monitoringthefuture.org/data/2002data-drugs.

The Gallup Organization: www.gallup.com.

Insurance Institute for Highway Safety (crash tests): www.hwysafety.org/default.htm.

Consumer Reports: www.consumerreports.org/main/home.jsp.

Good Housekeeping Institute (seal of approval): http://magazines.ivillage.com/goodhousekeeping.

Chapter 15

Dr. Ruth: *Family Circle,* 2/1/97 page 102. "Full Circle."

Adderall: *Woman's Day,* Feb 11, 2003, Advertisement following page 110. (Shire U.S. Inc.)

Chapter 16

The Gallup Organization: www.gallup.com.

Gallup's Pollwatcher's Guide: www.gallup.com/Publications/pollwatcher.asp.

The Harris Poll: http://vr.harrispollonline.com/register.

Zogby International: www.zogby.com.

American Medical Association: www.ama-assn.org.

National Rifle Association: www.nra.org.

U.S. Census Bureau: www.census.gov.

CBS celebrity activism survey: www.cbsnews.com/stories/2003/03/06/eveningnews/main543046.shtml.

CBS Internet dating survey: www.cbsnews.com/stories/2003/02/17/opinion/polls/main540870.shtml.

CBS pain survey: `www.cbsnews.com/stories/2003/01/28/ opinion/polls/main538259.shtml`.

Health surfing on the Web: `www.cnn.com/2000/HEALTH/08/18/water cooler.ap/`.

Investor optimism: `www.gallup.com/poll/releases/pr030224.asp`.

Worst cars of the millennium: `http://cartalk.cars.com/About/ Worst-Cars/results5.html`.

Riding with drunk drivers: `www.reuters.com/newsArticle.jhtml?type= healthNews&storyID=2345367`.

Children's health care: `http://my.webmd.com/content/article/60/ 67060.htm?lastselectedguid={5FE84E90-BC77-4056-A91C- 9531713CA348}`.

Crime Victimization Survey: `www.ojp.usdoj.gov/bjs/pub/pdf/cvus01.pdf`.

Breast Cancer: `http://my.webmd.com/content/article/36/1728_50583. htm?lastselectedguid={5FE84E90-BC77-4056-A91C-9531713CA348}`.

Cellphone use: `www.consumerreports.org/main/detailv2.jsp? CONTENT%3C%3Ecnt_id=23371&FOLDER%3C%3Efolder_id= 23051&bmUID=1047262402709`.

Cyber crime: `www.gocsi.com/press/20020407.html`.

Sexual harassment: `http://womensissues.about.com/library/ blsexharassmentstats.htm`.

Telling lies: *Journal of Applied Social Psychology,* 1997, v. 27, 12 pp. 1048-1062, by Ester Backbier, Johan Hoogstraten, and Katharina Meerum Terwogt-Kouwenhoven.

Chapter 17

National Cancer Institute: `http://cancer.gov`.

Marijuana and chemotherapy: *The New York Times,* September 16, 1975.

Information on clinical trials, U.S. National Institutes of Health: `www. ClinicalTrials.gov`.

HIV placebo: *The Manhattan Mercury* (Manhattan, KS), September 21, 1997.

Study linking older mothers and long life: *Kansas City Star,* September 11, 1997.

Prevalence of teen drug use: "Monitoring the Future," sponsored by the National Institute of Drug Abuse, and conducted by the University of Michigan. `http://monitoringthefuture.org/data/2002data-drugs`.

Chapter 18

National Institutes of Health (NIH): `http://www.nih.gov`.

Television Watching and Other Sedentary Behaviors in Relation to Risk of Obesity and Type 2 Diabetes Mellitus in Women: *Journal of the American Medical Association* (2003) Volume 289, pp. 1785-1791, (NIH News Release).

Anger expression inversely related to heart attack and stroke: *Psychosomatic Medicine* (2003), 65(1), pp. 100-110, (NIH News Release).

Light to moderate drinking and heart attack reduction in men: `www.nih.gov/news/pr/jan2003/niaaa-08.htm`, (NIH News Release).

Early treatment deters glaucoma: `www.nih.gov/news/pr/oct2002/nei-14.htm`, (NIH News Release).

Zinc for colds isn't a good idea: "Healthy Habits — that Aren't," *Woman's Day,* February 11, 2003.

Crickets and temperature:

> Journal article, *Garden Gate,* Issue Number 5: `www.gardengatemagazine.com/tips/25tip13.html`.

> "Cricket Chirp Converter": National Weather Service Forecast Office, `www.srh.noaa.gov/elp/wxcalc/cricketconvert.html`.

> Cricket chirp data: many different data sets exist for this example; here's the source most statisticians use: Pierce, George W., *The Songs of the Insects,* (1949), Harvard University Press, pp. 12–21. Note: I used only a subset of this data for my example (for illustrative purposes only).

> For more discussion on crickets and temperature (and to hear crickets chirping while you read): `www.dartmouth.edu/~genchem/0102/spring/6winn/cricket.html`.

Aspirin prevents polyps in colon cancer patients: National Cancer Institute's Cancer and Leukemia Group, led by Electra Paskett, The Ohio State University www.osu.edu/researchnews/archive/aspirin.htm.

Ice cream and murders (New York City): a good article to start investigating this subject and related issues is Spellman, B. A., & Mandel, D. R. (2003). For more on the psychology of causal reasoning, see Nadel, L. (Ed.) *Encyclopedia of Cognitive Science* (Vol. 1, pp. 461-466).

Chapter 19

U.S. Food and Drug Administration (FDA): www.fda.gov.

Total Quality Management (TQM): www.6sigma.us.

W. Edwards Deming: *Out of the Crisis* (Center for Advanced Engineering Study, Massachusetts Institute of Technology, 1986) www.deming.org.

Toothpaste tube filling machines: www.packagingdigest.com/articles/199710/52.html.

Tube-filling FAQ (frequently asked questions) source: www.keyinternational.com/FAQ_Tube_Filling.html#Q4.

Tube Council: www.tube.org.

Index

• *X* •

• *y* •

• *Z* •

Notes

Notes

Notes

Notes

Notes

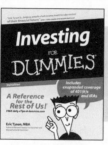

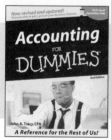

FOR DUMMIES®

A world of resources to help you grow

HOME, GARDEN & HOBBIES

0-7645-5295-3

0-7645-5130-2

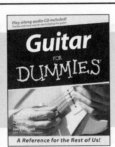

0-7645-5106-X

Also available:

Auto Repair For Dummies
(0-7645-5089-6)

Chess For Dummies
(0-7645-5003-9)

Home Maintenance For
Dummies
(0-7645-5215-5)

Organizing For Dummies
(0-7645-5300-3)

Piano For Dummies
(0-7645-5105-1)

Poker For Dummies
(0-7645-5232-5)

Quilting For Dummies
(0-7645-5118-3)

Rock Guitar For Dummies
(0-7645-5356-9)

Roses For Dummies
(0-7645-5202-3)

Sewing For Dummies
(0-7645-5137-X)

FOOD & WINE

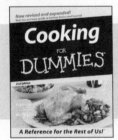

0-7645-5250-3

0-7645-5390-9

0-7645-5114-0

Also available:

Bartending For Dummies
(0-7645-5051-9)

Chinese Cooking For
Dummies
(0-7645-5247-3)

Christmas Cooking For
Dummies
(0-7645-5407-7)

Diabetes Cookbook For
Dummies
(0-7645-5230-9)

Grilling For Dummies
(0-7645-5076-4)

Low-Fat Cooking For
Dummies
(0-7645-5035-7)

Slow Cookers For Dummies
(0-7645-5240-6)

TRAVEL

0-7645-5453-0

0-7645-5438-7

0-7645-5448-4

Also available:

America's National Parks For
Dummies
(0-7645-6204-5)

Caribbean For Dummies
(0-7645-5445-X)

Cruise Vacations For
Dummies 2003
(0-7645-5459-X)

Europe For Dummies
(0-7645-5456-5)

Ireland For Dummies
(0-7645-6199-5)

France For Dummies
(0-7645-6292-4)

London For Dummies
(0-7645-5416-6)

Mexico's Beach Resorts For
Dummies
(0-7645-6262-2)

Paris For Dummies
(0-7645-5494-8)

RV Vacations For Dummies
(0-7645-5443-3)

Walt Disney World & Orlando
For Dummies
(0-7645-5444-1)

Available wherever books are sold. Go to www.dummies.com or call 1-877-762-2974 to order direct.

FOR DUMMIES®

Plain-English solutions for everyday challenges

COMPUTER BASICS

0-7645-0838-5

0-7645-1663-9

0-7645-1548-9

Also available:

PCs All-in-One Desk
Reference For Dummies
(0-7645-0791-5)

Pocket PC For Dummies
(0-7645-1640-X)

Treo and Visor For Dummies
(0-7645-1673-6)

Troubleshooting Your PC For
Dummies
(0-7645-1669-8)

Upgrading & Fixing PCs For
Dummies
(0-7645-1665-5)

Windows XP For Dummies
(0-7645-0893-8)

Windows XP For Dummies
Quick Reference
(0-7645-0897-0)

BUSINESS SOFTWARE

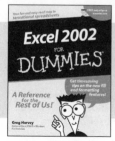

0-7645-0822-9

0-7645-0839-3

0-7645-0819-9

Also available:

Excel Data Analysis For
Dummies
(0-7645-1661-2)

Excel 2002 All-in-One Desk
Reference For Dummies
(0-7645-1794-5)

Excel 2002 For Dummies
Quick Reference
(0-7645-0829-6)

GoldMine "X" For Dummies
(0-7645-0845-8)

Microsoft CRM For Dummies
(0-7645-1698-1)

Microsoft Project 2002 For
Dummies
(0-7645-1628-0)

Office XP For Dummies
(0-7645-0830-X)

Outlook 2002 For Dummies
(0-7645-0828-8)

Get smart! Visit www.dummies.com

- **Find listings of even more *For Dummies* titles**

- **Browse online articles**

- **Sign up for Dummies eTips™**

- **Check out *For Dummies* fitness videos and other products**

- **Order from our online bookstore**

Available wherever books are sold. Go to www.dummies.com or call 1-877-762-2974 to order direct.

FOR DUMMIES®

Helping you expand your horizons and realize your potential

INTERNET

0-7645-0894-6

0-7645-1659-0

0-7645-1642-6

Also available:

America Online 7.0 For Dummies
(0-7645-1624-8)

Genealogy Online For Dummies
(0-7645-0807-5)

The Internet All-in-One Desk Reference For Dummies
(0-7645-1659-0)

Internet Explorer 6 For Dummies
(0-7645-1344-3)

The Internet For Dummies Quick Reference
(0-7645-1645-0)

Internet Privacy For Dummies
(0-7645-0846-6)

Researching Online For Dummies
(0-7645-0546-7)

Starting an Online Business For Dummies
(0-7645-1655-8)

DIGITAL MEDIA

0-7645-1664-7

0-7645-1675-2

0-7645-0806-7

Also available:

CD and DVD Recording For Dummies
(0-7645-1627-2)

Digital Photography All-in-One Desk Reference For Dummies
(0-7645-1800-3)

Digital Photography For Dummies Quick Reference
(0-7645-0750-8)

Home Recording for Musicians For Dummies
(0-7645-1634-5)

MP3 For Dummies
(0-7645-0858-X)

Paint Shop Pro "X" For Dummies
(0-7645-2440-2)

Photo Retouching & Restoration For Dummies
(0-7645-1662-0)

Scanners For Dummies
(0-7645-0783-4)

GRAPHICS

0-7645-0817-2

0-7645-1651-5

0-7645-0895-4

Also available:

Adobe Acrobat 5 PDF For Dummies
(0-7645-1652-3)

Fireworks 4 For Dummies
(0-7645-0804-0)

Illustrator 10 For Dummies
(0-7645-3636-2)

QuarkXPress 5 For Dummies
(0-7645-0643-9)

Visio 2000 For Dummies
(0-7645-0635-8)

Available wherever books are sold. Go to www.dummies.com or call 1-877-762-2974 to order direct.

FOR DUMMIES®

The advice and explanations you need to succeed

SELF-HELP, SPIRITUALITY & RELIGION

0-7645-5302-X

0-7645-5418-2

0-7645-5264-3

Also available:

The Bible For Dummies
(0-7645-5296-1)

Buddhism For Dummies
(0-7645-5359-3)

Christian Prayer For Dummies
(0-7645-5500-6)

Dating For Dummies
(0-7645-5072-1)

Judaism For Dummies
(0-7645-5299-6)

Potty Training For Dummies
(0-7645-5417-4)

Pregnancy For Dummies
(0-7645-5074-8)

Rekindling Romance For Dummies
(0-7645-5303-8)

Spirituality For Dummies
(0-7645-5298-8)

Weddings For Dummies
(0-7645-5055-1)

PETS

0-7645-5255-4

0-7645-5286-4

0-7645-5275-9

Also available:

Labrador Retrievers For Dummies
(0-7645-5281-3)

Aquariums For Dummies
(0-7645-5156-6)

Birds For Dummies
(0-7645-5139-6)

Dogs For Dummies
(0-7645-5274-0)

Ferrets For Dummies
(0-7645-5259-7)

German Shepherds For Dummies
(0-7645-5280-5)

Golden Retrievers For Dummies
(0-7645-5267-8)

Horses For Dummies
(0-7645-5138-8)

Jack Russell Terriers For Dummies
(0-7645-5268-6)

Puppies Raising & Training Diary For Dummies
(0-7645-0876-8)

EDUCATION & TEST PREPARATION

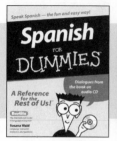
0-7645-5194-9

0-7645-5325-9

0-7645-5210-4

Also available:

Chemistry For Dummies
(0-7645-5430-1)

English Grammar For Dummies
(0-7645-5322-4)

French For Dummies
(0-7645-5193-0)

The GMAT For Dummies
(0-7645-5251-1)

Inglés Para Dummies
(0-7645-5427-1)

Italian For Dummies
(0-7645-5196-5)

Research Papers For Dummies
(0-7645-5426-3)

The SAT I For Dummies
(0-7645-5472-7)

U.S. History For Dummies
(0-7645-5249-X)

World History For Dummies
(0-7645-5242-2)